# Mathematics and Physics for Nanotechnology

# Mathematics and Physics for Nanotechnology

## Technical Tools and Modelling

**Paolo Di Sia**

CRC Press
Taylor & Francis Group
Boca Raton London New York

CRC Press is an imprint of the
Taylor & Francis Group, an **informa** business

First published 2019 by Pan Stanford Publishing Pte. Ltd.

Published 2022 by CRC Press
Taylor & Francis Group
6000 Broken Sound Parkway NW, Suite 300
Boca Raton, FL 33487-2742



ISBN-13: 978-981-4800-02-0 (hbk)
ISBN-13: 978-0-429-02775-8 (ebk)



**British Library Cataloguing-in-Publication Data**
A catalogue record for this book is available from the British Library.

**Mathematics and Physics for Nanotechnology:**
**Technical Tools and Modelling**

# Contents

# Preface

Nanobiotechnology is a new interdisciplinary science, with revolutionary perspectives arising from the fact that at nanosize the behaviour and characteristics of matter change with respect to ordinary macroscopic dimensions. Nanotechnology is a new way for producing and getting materials, structures and devices with properties and functionalities greatly improved or completely new.

The book provides an overview of the nanobiotechnology world along with a general technical framework of mathematical modelling, through which we study today the phenomena of charge transport at nanometre level. It also introduces new advances in analytical nano-modelling and discusses plasmonics and applications related to graphene. For these reasons, it is ideal both for introductory/specialised utilisation and as textbook/review book, a starting point for further research.

The book contains many solved exercises and practice exercises (with solutions), examples, illustrations and figures. It has been structured in such a manner that individual chapters can also be studied independently, for example, Chapter 1 is devoted to nano-education, Chapters 2 to 12 to mathematical and physical support to nanotechnology and Chapters 13 to 15 to mathematical modelling, plasmonics and graphene.

The book is not a purely mathematics or physics book, but it introduces the basic mathematical and physical notions because they are important and necessary for theory and applications in nanobiotechnology. It can also be considered as an extended formulary of the basic and advanced concepts.

The idea of writing this book was born from my desire to present an overview of analytical nano-modelling with a useful introductory basis of mathematics and physics. Following my participation to an international congress of nanotechnology in the United States, Mr. Stanford Chong offered me the opportunity to write the book with his prestigious publishing house. Sincere acknowledgments therefore go to him and to all his staff, who followed the different phases of creation and realisation of the book with attention, ability and competence.

The book will be useful to the greatest number of people and that it can be a starting point for discussions and insights as well as further development of the mathematical–physical modelling linked to the nanobiotechnology world.

I dedicate the book to all those who follow their ideas in life and pursue their choices with determination and firmness, in a free and independent way.

**Paolo Di Sia**
Autumn 2018

## Chapter 1

# Introduction

## 1.1 The Nanotechnologies World

The term 'nanotechnology' indicates a multidisciplinary approach concerning materials, devices and systems in which at least one of the three characteristic dimensions of their components is measured at nanometric scale (nm), i.e., the billionth part of metre: 1 nm = $10^{-9}$ m.

The nanometric scale characterises:

(a) The atomic sizes, with atomic diameters going from 0.1 nm (He) to 0.67 nm (Cs) (computed using quantum mechanical calculations).

(b) The molecular dimensions (proteins have typical extension from 1–20 nm).

(c) The atoms' distance at condensed matter level (the distance among sodium and chlorine ions in sodium chloride is 0.28 nm), is upto order of 100 nm for the smallest used components of microelectronics at the end of the twentieth century.

The submultiples of nanometre in the atomic world are more commonly expressed in angstrom (Å), with 1 Å = 0.1 nm = $10^{-10}$ m. The materials, whose structural and functional properties depend on components with at least one dimension at nanometric scale, are

*Mathematics and Physics for Nanotechnology: Technical Tools and Modelling*
Paolo Di Sia

ISBN 978-981-4800-02-0 (Hardcover), 978-0-429-02775-8 (eBook)
www.panstanford.com

said *nanostructural materials* and the nanometric components are said *nanostructures*.

The nanotechnologies control and manipulate the matter at nanometric scale and try to use the properties and the chemical–physical phenomena appeared at such scale. Many empirical technologies of the past years have been totally or partially understood and reconsidered in relation to structures and mechanisms at nanolevel, as ceramics, metallurgy, photographic process, heterogeneous catalysis, resins and polymers and special compounds for pneumatic tyres.

Nanophysics is commonly referred to as an intuition of Richard Feynman; in a famous conference of December 1959 by California Institute of Technology, he made forecasts around the possibility to control the matter and to realise devices at atomic scale, anticipating a great fields variety of scientific research and technical applications, that currently appear to be well developed. About that, we remember manufacture methods based on electronic and/or atomic beams, nanometric lithography, electronic microscopy, 'single atom' manipulation, electronics based on quantum and spin transport (so-called *spintronics*) and micro- and nano-opto-electro-mechanical systems, said respectively MOEMS and NOEMS. One of the first great step in such direction refers to the Japanese physicist Leo Esaki with the first realisation of a super network through a sequence of nanometric layers of different semiconductor materials (1969), opening the way to nanoelectronics. Eric Drexler of Massachusetts Institute of Technology (1977) put the experimental and computational basis of the conceptual and operational development for a lot of nanotechnologies. They assumed an important role in the succeeding 'information era', bringing to a scientific revolution comparable to the 'microscale' science and technology starting by 1970 (Fig. 1.1).

Electronics gave a strong impulse to the development of nanotechnologies through the progress in Physics, by which it depends on the fundamental understanding of processes at nanometric scale. The necessity to insert an always increasing number of electronic components in small volumes is due to both transportability and manageability demands of devices and also to the rapidity of calculation. Travelling electromagnetic signals at a finite speed (in 1 ns, the crossed distance is of the order of 30 cm), it is necessary that the central processing unit (CPU) of a calculator, able to perform 1 billion operations per second (1 Gflop), has smaller dimension than this length, so that only a negligible part of the calculation time is spent for the signal transmission to the components (Fig. 1.2).

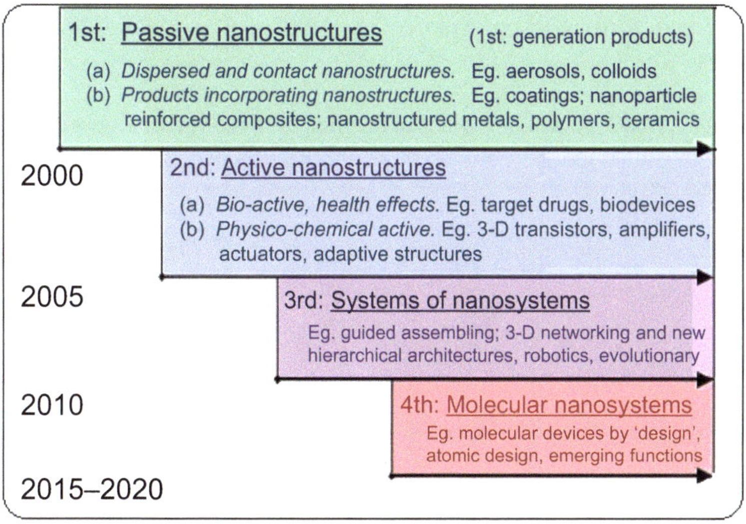


**Figure 1.1** Evolution of nanostructures and nanosystems.

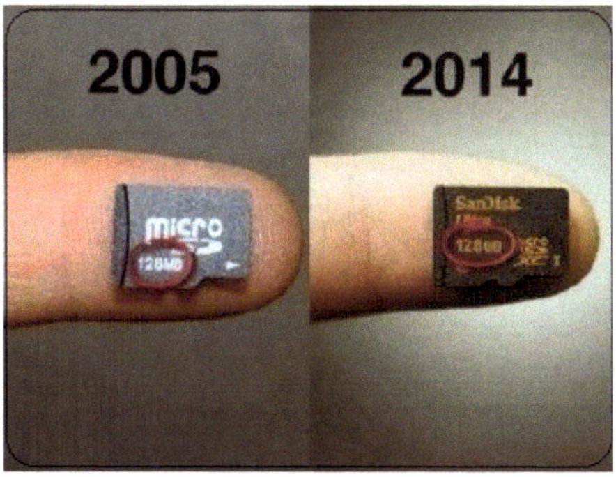


**Figure 1.2** Effect of miniaturisation: from 128 MB to 128 GB in the same space.

Being the number of placed electronic components of the order of millions, the respective dimensions have subsequently gone down (of the order of micrometre, μm). At such scale, electronics is commonly said microelectronics and it was developed through the invention of transistors (1947) and integrated circuits, in which the elementary components and the relative interconnections are realised on a single small plate (chip) of semiconductor material.

The number of active circuital elements in area unity is grown from $10^3$ elements/cm$^2$ (end of 1960) to $10^9$ elements/cm$^2$ and more, according to an exponential law known as *Moore's law* (Fig. 1.3).

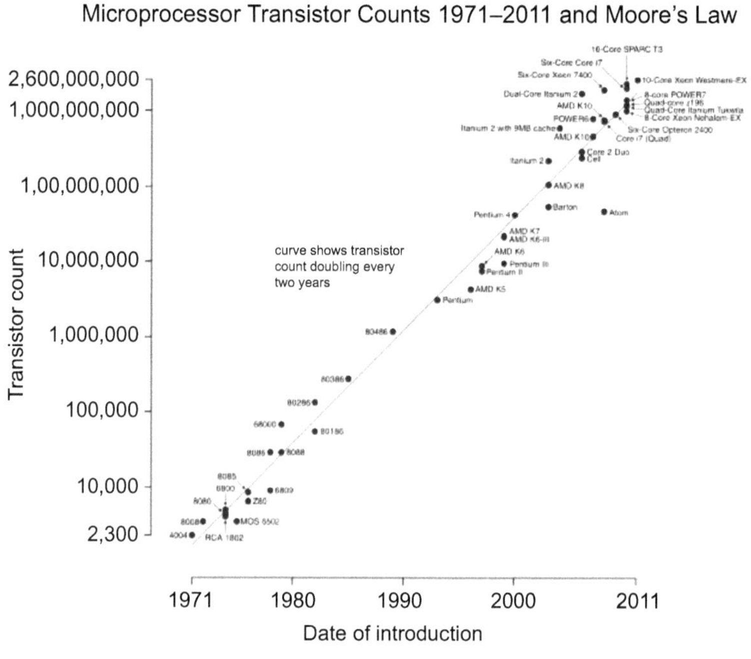


**Figure 1.3** The Moore's law.

The computer industry has kept on pushing the limits of miniaturisation and many current electronic devices have nanofeatures, whose origin is located in the computer industry, as compact disc (CD) and digital optical disc (DVD) players, cameras, inkjet printers, car airbag pressure sensors, etc.

Historically the Moore's law has been followed and maintained by the microelectronics industry for three principal reasons:

(a) *Economic reasons*: The cost of a chip is essentially tied to the silicon occupied surface, so at parity of cost, the smallest transistors are more functional.

(b) *Speed of work*: Reducing the transistors dimension, the inner current increases and contemporarily the quantity of charge required for the gate for turning ON or OFF is reduced.

(c) *Dissipated power*: Reducing the device's dimensions, reduces the number of necessary electrons for elaborating information, reducing consequently the dissipated energy for every calculation operation.

The integration level reached at the beginning of the twenty-first century entered in the nanometric domain, since the dimension of the smallest circuital elements used nowadays is broadly less than 100 nm. The distance going from the millimetric scale to 100 nm signs the integration levels, said respectively small-scale integrated (SSI), medium-scale integrated (MSI), large-scale integrated (LSI), very large-scale integrated (VLSI) and ultra large-scale integrated (ULSI). The integration of the components follows an 'exponential-type' growth too; it starts from multichip modules and becomes subsequently more rapid (by 2004) with the introduction of system-in-package (SIP) nanotechnologies and system-on-package (SOP) (Fig. 1.4).

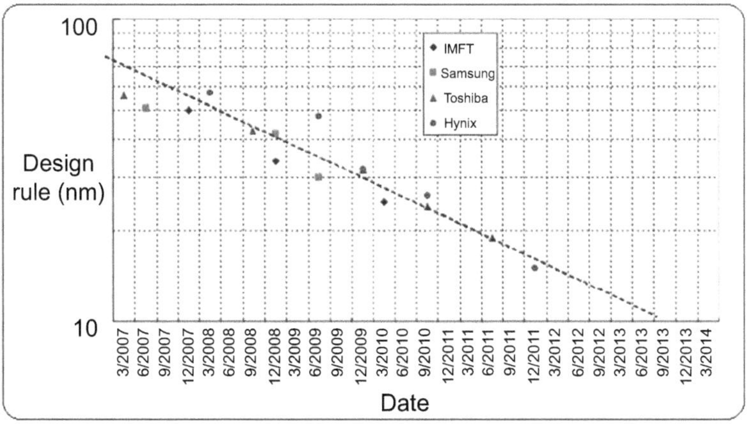


**Figure 1.4** The trend of scaling for NAND flash memory allows doubling of components manufactured in the same wafer area in less than 18 months.

At the atomic scale, the physics of the electronic transport processes in semiconductors changes radically; the components and the interconnections become smaller with respect to two characteristic lengths:

(a) *The electrons mean displacement*: At such distances, it holds a motion without collisions from ordinarily 'dissipative ohmic-type', the transport regime becomes ballistic.

(b) *The de Broglie wavelength of the electrons*: It brings to the appearance of quantum effects as the discretisation of the energy levels.

In the zero-dimensional (0-D) nanostructures, such as atomic clusters, nanocrystals and quantum dots, the discrete energy levels move in an appreciable way through the addition or the subtraction of a single electric charge; it is possible to control the tunnel effect through the passage of a single electron (Coulomb blocking of tunnelling) controlling the transport of single carriers. This produces a conceptual revolution in the sector of electronics and a limit to the validity of Moore's law in the present formulation. Another discriminating characteristic length among micro- and nanophysics is the wavelength of the electromagnetic waves at optical frequencies, corresponding to electronic energies of the order of 1 eV. It concerns the optoelectronic applications and the technical problems of microscopic observation, manufacture and checked manipulation of nanostructures. The observation and manipulation of matter at atomic scale have been possible through the invention of atomic force microscope (AFM), for its utility in the visualisation of non-conductive samples and for the manipulation at nanoscale, through a scanning tunnelling microscope (STM), coming to the great reached resolution of transmission electron microscope (TEM), laser scanning confocal microscopy (LSCM) and near-field scanning optical microscopy (NSOM). Such devices and techniques allow to visualise smaller objects than the wavelength of the utilised light. Many applications of nanotechnology range from simple to complex, as nanocoatings that can repel the dirt and reduce the need of dangerous cleaning agents, or mobile phones that are becoming smaller, cleverer and faster.

Nanotechnology, as 'engineering at a very small scale', can be applied to many areas of research and development, from medicine, manufacturing and computing to textiles and cosmetics, with a great impact on the everyday objects. It is providing solutions to many long-standing social, medical and environmental problems, in multidisciplinary research areas. For this reason, in particular at the phenomenological level, nanotechnology requires team efforts, including life scientists, biologists, biochemists, physicists, mathematicians, chemists, information technology experts and others.

In medicine, the improved knowledge of the body functioning at the cellular level is leading to new and better medical techniques, as the new 'body friendly' generation of implants, with a nanoscale topography that encourages acceptance by the cells in their surroundings (Fig. 1.5).

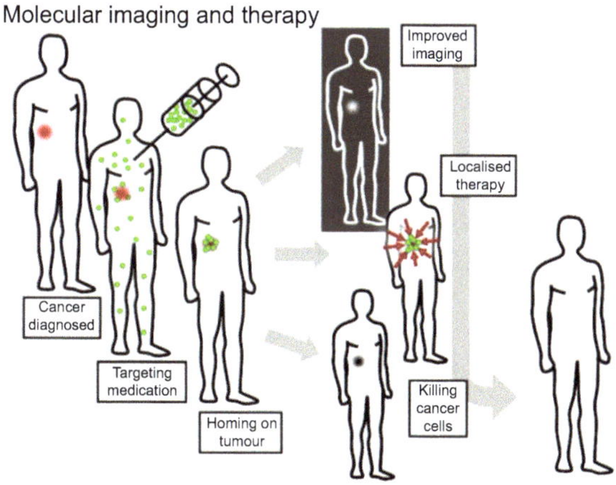


**Figure 1.5** Schematic illustration showing how nanoparticles or other types of drugs can be used for the treatment of cancer.

The applications of nanotechnology for new materials are very important, for example, in the coatings sector; we know that polymer coatings are manifestly easily damaged and affected by heat. Adding only a small percentage (of the order of 2–3%) of nanoparticulate clay minerals to a polymer coating, they become more durable and scratch resistant. This implies the possibility to use these materials in particular situations of protection from an external potentially corrosive environment.

Particular nanocoatings can prevent the adherence of pictures and 'graffiti', allowing an easy removal with water once the coating has been applied, having therefore the important effect of improving the urban environment.

As nanoscale particles are below the wavelength of visible light, therefore, they can offer new properties with a big range of possible applications, such as fluorescent nanoparticles and quantum dots. They can be made to exhibit a range of colours, depending on their size and composition.

As for all scientific sectors, it is important to work seriously and carefully with nanomaterials; although they offer enormous opportunities, nanomaterials may also have risks to be taken into account for the realisation of the full benefits. There is the possibility that free nanoparticles of a specific length scale may be harmful to health, for example, if inhaled, particularly at the manufacturing stage. Industry and government must be conscious of this fact and must focus on identifying particles that may be hazardous to health or environment and studying how these risks may be quantified and minimised. In every case, nanotechnology has doubtless a large potential to bring benefits to society, but it must ensure these advances are as safe as possible.

## 1.2 Classification of Nanostructures

Two opposite procedures of manufacture of nanostructures are possible, i.e., 'top-down' and 'bottom-up'. With the 'top-down' procedure, the obtained structures preserve the chemical–physical qualities of the started solids. In the 'bottom-up' procedure, it is possible to realise a great variety of nanostructured materials, also not existing in Nature, with non-ordinary properties and different from those of the usual crystalline materials of equal chemical composition (Fig. 1.6).

The small atomic or molecular clusters, composed of a number of atoms or molecules going from unities to a few thousands, form structures that can result very different by those of the respective solids and consequently their properties will be different; for example, small gold clusters have catalytic activity, while the crystalline material is notoriously inactive; the silicon clusters can be luminescent, while the crystalline silicon does not have this peculiarity. One of the key ideas of the 'bottom-up' procedure is the use of 'pre-formed' clusters as bricks for the construction of nanostructured materials, with the purpose to preserve on a macroscopic scale particular properties, typical of the clusters. Therefore, it becomes possible the realisation of materials with desired properties and high performance through the choice and control of their nanometric constituents. The combination of 'top-down' methods and tools with 'self-assembly' processes at atomic scale offers considerable and innovative opportunities to unite chemistry and biology parts to artificial structures created by man.

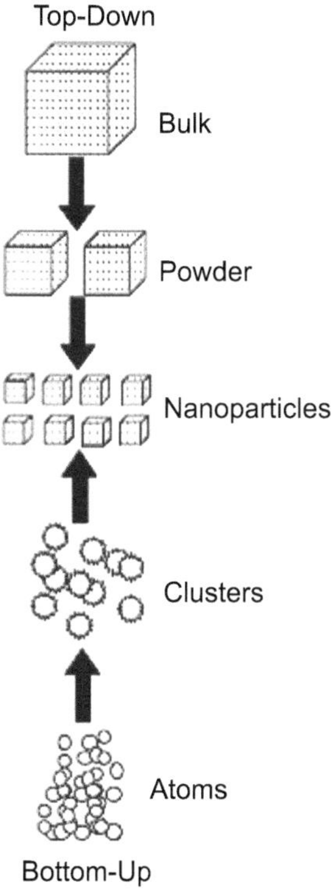


**Figure 1.6** Schematic representation of the 'top-down' and 'bottom-up' procedures.

The manufacture procedures of nanostructured materials and systems depend on the nature of the 'ground' elements of nanometric dimension composing them; therefore, the nanostructures can be classified in relation to the respective elementary components. In particular:

(a) *Atoms, inorganic molecules and metal-organic compounds*: The first artificial nanostructures, realised in the last decades of 1900, were grown on a high regular surface of a solid substrate as a single ultra-thin films, quantum wells or multilayers formed by different materials, characterised by constant composition and thickness and totally controllable at atomic scale. The deposition techniques are based on the transport in the growth region of atoms or molecules:

   (i) Either free (direct) adsorption in vapour state or transportation by chemical mixtures is done and then released to the surface through chemical deposition from vapour state or chemical vapour deposition (CVD), respectively.

   (ii) In an ejection created by the checked evaporation of a material (sputtering), or from a sub- or super-sonic beam of atoms or molecules expelled by one or more temporally checked sources, a technique known as *molecular beam epitaxy* (MBE). Microelectronics, optoelectronics, the high number of 'plane technology' devices and the most recent 3-D technology advanced through the development of 'molecular epitaxy' techniques.

(b) *Organic molecules, polymers, dendrimers and biomolecules*: In previous years, there was a great progress in the preparation of organic objects able to be assembled in structures with increasing complexity. An essential contribution derived from 'supra-molecular chemistry'. The discovery of polymers with new topological forms, said *dendrimers*, has led to a new class of nanometric components with particular functional properties, suitable to special coverings and to the formation of biocompatible buffer layers. It has been used with success nanoparticles composed of complexes among policationic lipids (or polymers) and deoxyribonucleic acid (DNA) in the transport and release of medicines and genes *in vivo*.

The 3-D periodic mesoscopic structures, realised by self-assembly of organic molecules in solution, can be used as stamps for the creation of high porous and periodic 3-D inorganic architectures on the wavelength scale of visible light (hundreds of nanometre), suitable to realise the so-called 'optical networks'. Such networks allow the realisation of elements of photonic circuits, in which the currents are replaced by light signals that can be driven, trapped, amplified and cancelled. The recent synthesis of metalorganic molecules incorporating magnetic atoms opened the way to the molecular nanomagnetism, with interesting perspectives for the development of ultra-dense magnetic memories.

(c) *Clusters, nanocrystals, nanoaggregates and quantum dots*: There are various techniques able to prepare current inorganic materials in the form of small atomic agglomerations, with dimensions going from few unities to some millions of atoms and with forms that can vary from the amorphous structure [few unities of atoms (clusters)] to the crystalline structure [for sufficiently great numbers of atoms (nanocrystals)]. The procedures allow to deposit quantum dots (also called *artificial atoms*) on substrates for the creation of nanostructured materials with checked structure and defined functionality, finalised to the use of particular mechanical, tribologic, chemical and electromagnetic properties. Amorphous and porous nanoaggregates can also be obtained through chemical erosion of crystalline material, with possible big changes of physical properties.

(d) *Fullerenes, nanotubes and carbon nanostructures*: With the identification by Robert Curl, Harold Kroto and Richard Smalley of the carbon clusters in the form of closed cages with covalent graphitic-type ($sp^2$) bonds, said 'fullerenes', a large research field of carbon nanostructures is born. The most frequent fullerene in Nature and in the ordinary synthesis processes has formula $C_{60}$ and forms a spherical-type molecule constituted by 12 pentagonal rings and 20 hexagonal rings (Fig. 1.7).

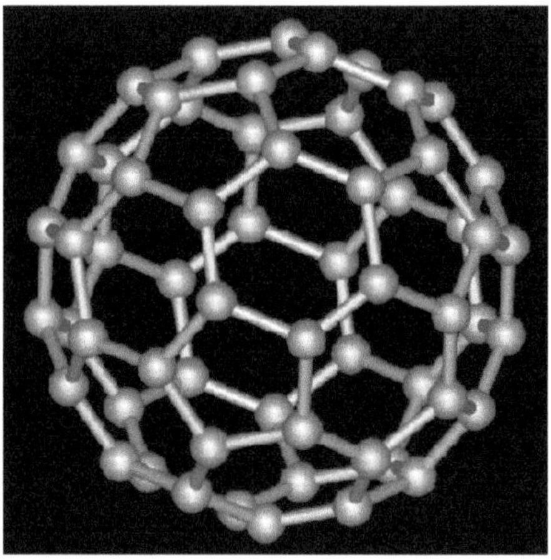

**Figure 1.7** Fullerene $C_{60}$.

The fullerenes research has brought to the discovery and the synthesis of carbon nanotubes with cylindrical structure, with nanometric diameter and length of greater orders too, as well as of the concentric fullerenes and nanotubes ('onions' and 'multiple wall nanotubes' respectively), of the fullerides of alkali metals and varied mixtures and fullerenes polymers with interesting optical, electrical and magnetic properties. Such new graphite-type carbon forms, said also 'graphenes', originate aggregated forms:

(i) The pure fullerite ($C_{60}$ solid), in which 0-D objects ($C_{60}$ clusters) are kept together in all spatial directions by van der Waals forces.

(ii) The nanotube bundles, where 1-D objects (the nanotubes) are kept together by van der Waals forces in two spatial directions.

(iii) The graphite, where 2-D objects (the graphitic plans) are kept together by van der Waals forces in the third spatial direction.

The 3-D graphenes have been also hypothesised, with form of minimal periodic surfaces said 'schwarzites'; it has been synthesised a new nanostructured carbon form, referable to an amorphous schwarzite form.

Carbon nanotubes are chemically inactive and have particular mechanical properties; such characteristics stimulated a great research activity in relation to practical applications, such as electrodes with very high specific surface for super-capacitors, field-effect electron emitters for ultra-flat screens and X-ray sources, molecules vectors of biological interest, electrodes for electrochemical characterisations at nanometric scale. About the electronic transport, carbon nanotubes behave as half-metals, similar to graphite, or as semiconductors, in relation to the orientation of the hexagonal rings with respect to the axis of the nanotube. Such properties, together with the possibility of nanometric manipulation, opened very promising sectors of application, as nanoelectronics and molecular electronics.

Carbon nanotubes have surprising structural, electronic and thermal properties. They are highly conductive both to electricity and heat, with an electrical conductivity as high as copper and a thermal conductivity as great as diamond. For these reasons, they offer significant possibilities for creating nanoelectronic devices, circuits and computers. Carbon nanotubes have also extraordinary mechanical properties, being 100 times stronger than steel, while only one-sixth of its weight. These mechanical properties offer enormous possibilities for the production of new stronger and lighter materials for medical applications, aerospace, environment, everyday life, etc. There are already products containing nanotubes in the market, as in some tennis racquets, where nanotubes are used to reinforce the frame and improve the ability to absorb shocks.

(e) *Nanotubes and nanowires of lamellar and metallic compounds*: There are many other compounds well able to form lamellar solids, similar to graphite. From an electric point of view, the polyatomic nanotubes reflect the properties of the started lamellar compounds and it is possible to have metals, or semiconductors, or insulators; the range of the possible technological applications based on nanotubes has notably expanded: near nanotubes are added nanowires.

The quantum charge transport is obtained equally well with metallic nanowires, with diameters of the order of few interatomic distances (order of 1 nm), with a structure checked at atomic level. The growth of an opportune substrate of nanometric arrays of electroluminescent material allows to realise a regular LED or semiconductor laser distribution, separated by micrometric distances and therefore suitable for the construction of pixel networks for ultra-flat screens. The nanostructured materials based on the assembly of nanometric constituents are characterised by ratios of the surface and the volume (order of hundreds or thousands of $m^2/g$); such geometric property makes them ideal for the use in composite materials, chemical reactions, heterogeneous catalysis, adsorption or desorption, release of medicines, super-capacitors, energy storage and other applications. It has been observed that the systematic organisation of the matter at nanometric scale is a determinant aspect of the biological systems, which are able to realise complex structures through self-assembly processes; such processes have been observed also in the non-biological inorganic and organic systems, in which they are realised under particular thermodynamic conditions. It is also highly studied the direct use of biomolecules (DNA, virus, bacteria) for the assembly of nanostructures. The programmable self-assembly properties of DNA are also opening various possibilities for the manufacture of mesoscopic functional structures and for molecular electronics. The organic self-assembled nanostructures serve also as moulds for the synthesis of ultra-porous materials of inorganic aerogel and nanocomposites for membranes and molecular filters, catalysts and matrixes for the textures reparation. The processes of self-replication of polymeric structures, already known to industrial level at micrometric scale, have been shown at nanometric scale too. The same living organisms constitute the most complex matter organisation currently known, being structured both on a nanometric scale (biomolecules) and on micrometric (cells), millimetric (textures) and metric scales (organs). The objective of the bottom-up nanobiotechnologies is also the manufacture of complex devices, characterised by an analogous multiscale organisation. The direct manipulation of atoms and molecules deposited on a substrate surface can be performed with scanning micro- and nanoprobes, based on the principle of the scanning electronic microscope and the atomic force microscope. It results possible to transport atoms and molecules along the substrate surface and to get atomic alignments, filaments, closed fences and more complex pattern, suitable for experiments of quantum transport, nanochemical applications and information storage. By the theoretical Physics viewpoint, the bottom-up procedure utilises the considerable progress of quantum mechanics of many body systems and the development of supercomputers that allow to simulate the dynamic evolution of electrons and atoms in wide systems, typically of the dimension of clusters composing the nanostructured materials. This allows a 'first principles' description, through the quantum calculation, of the elementary properties of the nanometric components. In this way, we have the welding among theoretical simulation activity and experimentation on real systems, in the past separated by considerable differences of length and time scales.

## 1.3 Applications of Nanotechnologies

A lot of industrial products based on nanotechnologies are currently available: from the new particularly resistant tyres to the use of medicines constituted by nanoparticles for a targeted and efficient release, from colours and pigments for the press (considerably improving the quality and stability) to semiconductor lasers and heads for reading or writing at high resolution and speed on optic and magnetic discs. Among the sectors that have and will have great contribution by nanotechnologies, we remember:

(a) *Car and aeronautics industries and exploration of the space*: These sectors cover the development of materials reinforced with nanoparticles for the light parts, pneumatics reinforced by nanoparticles, resistant to the usury and recyclable, anti-dust paints, not inflammable plastics with low costs, self-repairing coverings and textures, ultra-light space vehicles, economic management of the energy, small and efficient robotic systems, special protective coverings with high resistance to corrosion and erosion, thin layers for the optic filtration and thermal barriers, polymers and composite nanostructured materials (Fig. 1.8).

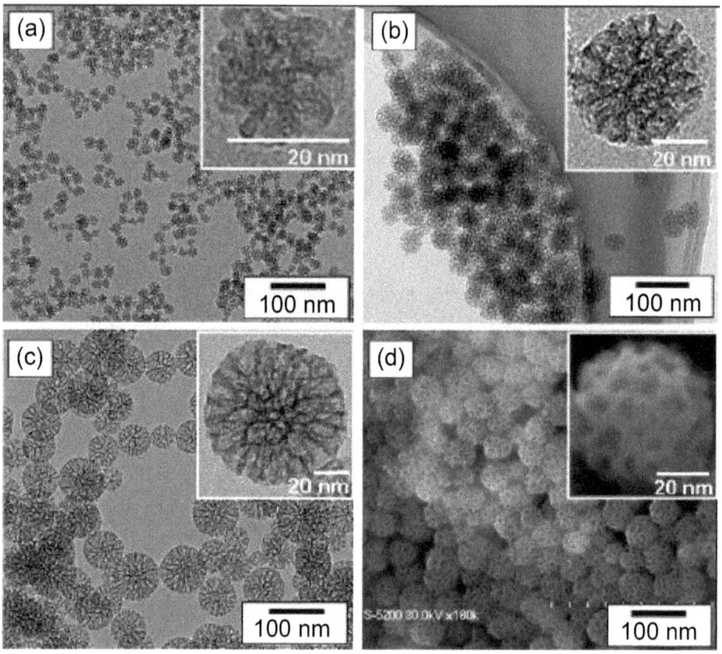


**Figure 1.8** Transmission electron microscopy (TEM) images (a, b and c) of prepared mesoporous silica nanoparticles with mean outer diameter: (a) 20 nm, (b) 45 nm, (c) 80 nm. Scanning electron microscope (SEM) image (d) corresponding to (b). The insets are high-magnification images of mesoporous silica particle.

In the cars sector, we remember also the improving of lubricants, the use of fuel cells for clean energy, lighter and stronger engine and body materials, better catalysts, nanoporous filters, self-cleaning windshields, self-healing and scratch-resistant coatings, environmentally friendly corrosion protection, the increase of their high technological contents, using smart nanosensors for the prevention of possible problems.

Important advances came also by the use of polymer nanocomposites for body panels, lightweight yet rugged and in new metal nanocomposites to improve engine efficiency. Particular designed nanoparticles are used as fuel additives to lower consumption in commercial vehicles and for the reduction of toxic emissions, with attention to be more environmentally friendly in the manufacturing processes as well as in the final product. Efforts are made to investigate the nanotechnology possibilities for reducing the toxic wastes by substituting new nanomaterials for hazardous reactants and solvents and by using nanotechniques to completely eliminate their need.

(b) *Electronics for communications, sensoristics and electromechanical systems*: These systems cover development of recording systems based on quantum nanostructures, ultra-flat screens, wireless technologies, new devices and processes in every technological sector of information and communication based on dating-storage ability and continuously increasing calculation speed, realisation of switches with single organic molecules and non-volatile memories with a bits density 1 million times greater than the current dynamic random access memory (DRAMs). At nanometric scale, there is tunnelling electrons phenomenon that brings the realisation of tunnel-resonant devices and Josephson junction systems with potentiality in quantum computation. Also spintronics, based on the transport of defined spin electrons, has high realisation perspectives at nanometric scale (Fig. 1.9).

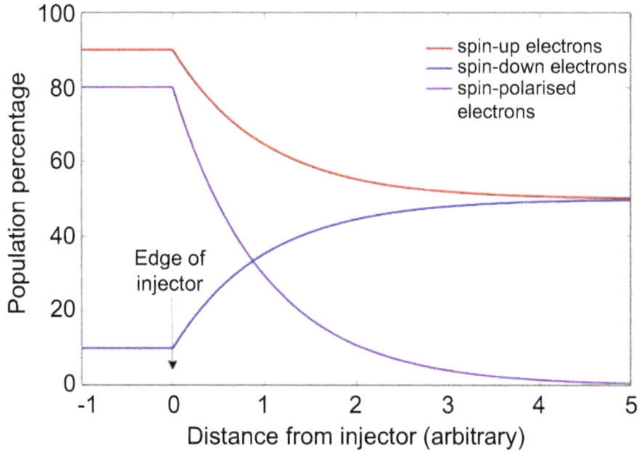


**Figure 1.9** The plot shows a spin-up, a spin-down and the resulting spin-polarised population of electrons. Inside a spin injector, the polarisation is constant, while outside the injector, the polarisation decays exponentially to zero as the spin up and down populations go to equilibrium.

(c) *Chemical products, materials for energetics, energy storage and contamination of the environment*: These cover the development of new battery types, artificial photosynthesis for the production of clean energy, 'quantum hole' solar cells, energy saving through the use of ultra-light materials, catalysts for the growth of the energy efficiency of chemical plants and reduction of the polluting emissions from the motor vehicles. Notable attention had also aerogels (Fig. 1.10), spongy highly porous materials endowed with 3-D nanostructured texture, very promising in the sector of catalysis and energy accumulation.

The conversion of solar energy into chemical energy has a strong impulse thanks to the photochemical Graetzel cells, consisting of nanostructured $TiO_2$ films, on which dye molecules adsorb.

**Figure 1.10** Aerogel.

Also the lithium batteries cathodes and the components of the combustion cells with the correspondent hydrogen storage, use nanostructured materials. The low-cost synthesis of nanoporous zeolites led to a revolution in the catalysis of petrochemical processes.

Nanotechnology is leading to improve better ways of storing energy, searching to reduce damages to the environment, in particular with toxic chemical materials, through the development of new nanocoatings and nanostructured surfaces, which can effectively repel dirt and other contaminants. Many anti-corrosion coatings involve chromium and cadmium, both deadly substances that the EU is seeking to limit. New smart nanocoatings, non-toxic and highly efficient, are in the process of development.

Heavy metals and other pollutants are thrown into the atmosphere by emission of industrial processes, producing serious contamination of the environment; most of these particles and gases can be re-claimed and re-used with specially functionalised nanomaterials, placed in the waste gas stream.

(d) *Pharmaceutical products, health protection and life sciences, biomedical applications and cosmetic sector* (Fig. 1.11): These cover the development of new nanostructured medicines, systems of medical release and genetic material for specific body parts, biocompatible prosthesis and suitable substitutes of physiological fluids, self-diagnosis tools, sensoristics for 'on-chip' biological tests, materials for the regeneration of bones and other tissues. We remember besides the numerous nanosystems and the devices able to determine the sequence of single DNA molecules, with great perspectives in the 'large-scale' genomics and the biological tissue engineering, which uses cells and their molecules for the creation of substitutes to bad or not-working tissues.

**Figure 1.11** Nanotechnology in cosmetics sector.

In the sector of health protection, for an earlier detection of diseases, research focuses on introducing into the body particular designed nanoparticles, composed of tiny fluorescent quantum dots for targeting anti-bodies, which bind to diseased cells. Quantum dots fluoresce brightly and this can be picked up by particular developed imaging systems, enabling the precise pinpointing of a disease, even at a very early stage.

Nanotechnology is also leading to faster diagnosis, normally lengthy and stressful, with results that can take several days or even weeks for coming. Many tests can be built into a single, often palm-sized device, requiring tiny quantities of sample; samples can be rapidly, almost instantaneously processed and analysed.

Traditional drugs may have unpleasant side effects, because the body needs to be flooded with high doses of a drug for ensuring that a sufficient volume reaches the site of the disease. Using special designed drug-carrying nanoparticles, much smaller quantities of a drug are necessary, therefore reducing the toxicity of the body. The drug can be activated only at the disease site (by light or other means) and the progress of the cure can also be monitored using advanced imaging techniques.

(e) *Manufacturing industry and textile sector*: These cover the development of the precision engineering based on measure techniques at nanometric scale, sinterised nanopowders for materials with specific properties, polishes with nanoparticles, self-assembly of structures from molecules, biostructures, biomimetic materials, realisation of dispersions at nanometric scale, as the use of colloidal silica nanoparticles in the manufacture of optic fibres through 'sol–gel' processes, plastics with inorganic nanoparticles based on delayed inflammability, surface coverings with nanoparticles for increasing the resistance to the usury and the chemical corrosion, self-lubrification, change of the optic properties, nanometric powders for air decontamination from harmful bacteria, nanoparticles for inks and dyes, ferrofluids obtained through the dispersion of ferromagnetic nanoparticles in a liquid, manufacture of structural materials with high performance through dispersion of nanometric particles in solid phases, in order to obtain steels, alloys, ceramic materials with increased hardness and increased mechanical properties, cutting instruments of extraordinary hardness and reduced fragility, ductile cements and medical implants able to be integrated with biological tissues.

New ideas and technologies are used in the textile industry, not only for the fashion conscious, but in particular in the aerospace, automotive, construction, healthcare and sportswear industries, through self-cleaning and stain repellent nanotechnologies and also for school clothes. The electrospinning techniques are using nanoparticles, enzymes, drugs or catalysts that can be anti-bacterial and wound healers.

Nano-electro-devices can be embedded into textiles to provide special support systems in dangerous professions or sports. Some garments can provide monitoring of inner temperature, chemical sensing and power generation for enabling communication with the outside world, vital for the safety of firefighters working in dangerous situations, or for skiers or rescuers, to give early warning signs of hypothermia.

Research is ongoing into artificial nanofibres, where clay minerals, carbon nanotubes and metal oxide nanoparticles are used to give new properties, providing flame retardancy for a fabric, increased strength and shock-absorbency, heat and UV radiation stability. Efforts are made in the area of inkjet printing onto textiles, opening many possibilities in creating flexible electronic and sensing materials.

Carbon nanotubes can also be mixed with many different materials such as plastics and textiles to produce lightweight bullet-proof vests. One of the greatest potential for creating new products lies in harnessing the electrical properties of lightweight and robust nanotubes to generate heat, with applications from electric blankets to heatable aircraft wings and wallpaper heating for cold walls.

(f) *Environment, safety and monitoring systems*: These cover the development of applications regarding the selective membranes for the contaminants filtration, nanostructured traps for pollutants removal from industrial outflows, reduction of pollution sources and growth of the recycling possibilities. In the safety field, we remember also detectors, detoxificators of chemical and biological agents, overseeing miniaturised security systems, mimetic materials, ultra-hard nanostructured materials and coverings and self-repaired tissues.

Nanotechnology offers exciting discoveries in environmental friendly technologies, as to extract renewable energy from the sun for the prevention of pollution. The planet absorbs our excesses, but with the ongoing destruction of the rainforest (responsible for more than 1/4 of carbon emissions), while the population of the world reached over 7.5 billion; the preservation of the planet is therefore a right-duty for us and for the future human generations, for limiting the damage and ensure a future for children. The fossil fuel is responsible for more than 40% of the carbon dioxide annually emitted; energy from sunlight is sufficient to meet the human needs 10000 times over. For these reasons, more efficient and cheaper solar energy collectors are actually in the process of being developed using nanotechnology. These collectors could be deployed as small units in private homes; they work well in diffuse light too, so would suit even less sunny climates, with a big benefit related to avoid the sterilisation of precious land and therefore improve the quality of people's lives. New cheap nanosensors are developing, for a fast, accurate and *in situ* pollution monitoring.

(g) *Food and drink*: Many foods and beverages contain natural components which are in the nanosize range; the manipulation of the occurring nanoparticles involved in processing as dairy products has been undertaken for some time. The improving understanding of mechanisms, like the targeted delivery, enabled food companies to deliver flavours, vitamins and minerals offering health benefits and giving new physical effects to foods. This leads to a big growth in the sector of nutraceuticals and functional foods, with the appearance of new food products with new tastes, flavours and textures. We remember also the applications of nanotechnology to food manufacturing anti-bacterial work surfaces, filters able to extract toxins, packaging providing a better barrier against contamination, with the signalisation by change of colour, when its content is spoiling.

Another interesting field for the application of nanotechnology is the encapsulation and delivery technologies for flavours and fragrances. Started for the delivery of pharmaceutical drugs, these technologies found new applications in foods and household products. Encapsulation is an important way to improve the attributes and performance of a less-than-stable substance that might be affected by air and/or light. With this technique, the active ingredients have a longer life, more stability and protection. It is a way of delivering enhanced taste and ensuring the right daily doses of vitamins and minerals. Moreover, nano-encapsulation techniques can provide long lasting scents in household fragrances; the slow release of enzymes and other agents in washing machines and dishwashers can help the reduction of energy and water use.

## 1.4 Applied Mathematics and Nanotechnology

During previous years, the essential techniques of theory, modelling and simulation developed similarly to the remarkable experimental progress on which the new field of nanoscience is based. This time, the development of algorithms, Monte Carlo quantum techniques, *ab initio* molecular dynamics, progress in Monte Carlo classical methods and mesoscale methods for soft matter are seen. Simultaneously, advances in computing hardware increased the computational power of many orders of magnitude. The combination of new theoretical methods with computational power made possible the simulation of systems with millions of freedom degrees.

The application of new and exciting tools to nanosystems created an urgent need for a quantitative understanding of matter at nanoscale. The limited number of quantitative models, which describe the observed phenomena in new ways, limit the progress in the field. Without new models for the quantitative description of the structure and dynamics at nanoscale, nanoscience proceeds with difficulty, especially in common sectors, as the molecular electronics and the biomolecular materials. For these reasons, it is important to emphasise the efforts for the creation of new predictive theoretical models, at numerical and analytical level.

Many important phenomena have interaction with their environments in the macro temporal and spatial scales, but should be designed as 'first principles' at atomic scale. The time scale required to solve quantum oscillations may be of the order of $10^{-15}$ s, while the protein folding requires microseconds and the engineered nanosystems may require even longer times of simulation. Analogously, the spatial scales can range from angstroms to centimetres. So, nanoscience poses unprecedented challenges to the mathematical representation and to the multiscale analysis. Models and mathematical algorithms must cover the range from discrete to continuous and from deterministic to random.

The use of fast algorithms can perform larger simulations (for example, to include in a model many atoms or basis functions) and can simulate physical processes in more detail and have computation with faster kernel to improve statistical samples.

With the use of linear algebra in electronic structure calculation, simulation methods based on the *ab initio* electronic structure are important in the present and future investigation of the properties of materials at nanoscale. The ability of these methods to predict electronic and structural properties without previous empirical knowledge and/or without an experimental input, makes them very attractive.

Monte Carlo methods are a class of stochastic optimisation techniques that are particularly reliable for the transition probabilities of statistical mechanics and quantum physics; they are therefore applied to many problems, not only to the commonly considered optimisation problems. The quantum Monte Carlo method is one of the most accurate methods for the electronic structure, extendable to systems in the nanorange. It numerically solves with great accuracy the Schrödinger equation. Kinetic Monte Carlo methods are an important class of techniques that deal with a range of atomic particles and aggregate phenomena, in addition to those of the electronic structure. Also the computational geometry is a mature area of computer science algorithmics.

Important investigation areas are as follows:

(a) Nanoconstituents (nanotubes, quantum dots, clusters, nanoparticles)

(b) Complex nanostructures and nano-interfaces

(c) Assembly and growth of nanostructures, in particular in relation to:

   (i) determination of the essential mechanisms of transport at nanoscale;

   (ii) realisation of theoretical and simulation models for the study of nano-interfaces, very important in complex and heterogeneous nanoscale systems;

   (iii) simulation with reasonable accuracy of the optical properties of nanoscale structures and modelling of optoelectronic devices;

   (iv) simulation of complex nanostructures involving 'soft' and 'hard' structures and nano-interfaces between 'hard' and 'soft' matter;

   (v) simulation of self-assembling;

   (vi) quantum coherence, decoherence and spintronics;

   (vii) development of self-validating methods.

The role of applied mathematics in these areas is fundamental, for formulating new theories and developing new computational algorithms, applicable to the complex systems at nanoscale.

The experimental techniques for the controlled fabrication of nanotubes and nanocrystals, quantum dots and wells, produced an entirely new set of elementary nanostructures. Combinatorial chemistry and genetic techniques opened to the synthesis of new biomolecular materials and the creation of nano-interfaces and nano-interconnections between soft and hard matter.

The density functional theory (DFT) transformed theoretical chemistry, surface science, physics of materials through a new way to describe the electronic structure and the interatomic forces in molecules with hundreds and thousands of atoms; the molecular dynamics, with fast multipole methods for computing long-range interatomic forces, made accurate calculations on the dynamics of millions and billions of atoms. Classical Monte Carlo methods for simulations have been expanded and quantum Monte Carlo methods promise to provide an almost exact description of the electronic structure of molecules. The Car–Parrinello method for *ab initio* molecular dynamics with simultaneous computation of the electronic wave functions and the interatomic forces opened the way to the exploration of the molecular dynamics in condensed media and complex interfaces. The growth of fast workstations, cluster computing, new generations of massively parallel computers, completes the picture of the transformation of theory, modelling and simulation of the last two decades. The study of structures, dynamics and properties of systems in which one or more spatial dimensions are at the nanolevel, results from distinctly different dynamics and properties, often in unexpected ways that can be favourably exploited by both systems with small molecules and by macroscopic systems in all dimensions.

We can consider the following three big classes of nanosystems:

(a) *Nanobuilding blocks*: Nanobuilding blocks such as nanotubes, quantum dots, clusters and nanoparticles can be sinterised in a completely reproducible way and experimentally well characterised. They are the central element of the new nano-mechanic, nanoelectronic and nanomagnetic devices. It is crucially a quantitative understanding of transport, dynamic, electrical, magnetic, thermodynamic and mechanical properties. It is believed that the best characterised and most appropriate building blocks are clusters, molecular nanostructures, nanotubes and related systems, wires, films and quantum dots, are well defined structures.

    About the transport at nanoscale, new quantum phenomena happen, such as tunnelling, fluctuations, confined dimensionality, discretisation of the electric charge. At these dimensions, the system is dominated by the surface, presenting further complications for electronic materials.

    About the optical properties at nanoscale with confined dimensions, the optical properties of matter are often altered. Silicon, as an example, has poor optical properties for optoelectronic devices, such as solar cells or lasers, but at small dimensions, properties dramatically change. Nanotechnology has also the potential to do much advances for medical technology and biosafety. The synthesis of nanostructures of soft and/or biological matter with traditional semiconductors provides new opportunities for the creation of new devices. A big number of applications, such as biosensors for national security, distribution and monitoring of drugs, discovery of disease, are reasonably obtainable through the research at nanoscale.

(b) *Complex nanostructures and nano-interfaces*: Central for nanoscience is the assembly and manipulation of the fundamental building blocks for creating structures, materials and functional devices (Fig. 1.12).

Nanostructured materials are complex and heterogeneous in form and substance, composed of different materials and dominated by nano-interfaces.

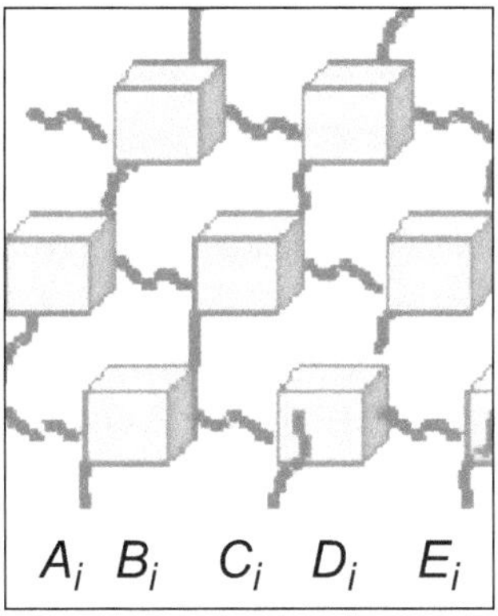


**Figure 1.12** Schematisation of an assembled structure of nanoscale building blocks, bound by organic paths.

It is crucial to understand how interfaces and complexity at nanoscale control properties and characteristics to larger scales. In spintronic devices, the nano-interfaces among ferromagnetic and anti-ferromagnetic domains can control the switching speed for faster computers. The large surface to volume ratio, due to the prevalence of nano-interfaces, combined with the complexity of nano-interfaces and nanostructures, gives various challenges in the development of predictive theories in nanoscience. For these reasons, engineered nanoscale materials are distinctly different from conventional bulk materials; in particular, when required for peculiar simulations and *ab initio* calculations.

(c) *Dynamics, assembly and growth of nanostructures*: Central aspects of this sector are transport properties (electronic, spin and molecular diffusion transport), dynamic processes leading to their creation and in particular the self-assembly. The focus results in time-dependent properties of nanostructures (such as transport properties) and time-dependent processes used to produce nanostructures and nanostructured materials (such as direct self-assembly, nucleation and growth and methods of vapour deposition). There are a wide variety of relevant transport mechanisms in nanoscience that include the electron transport (in molecular electronics, in nanotubes and nanowires), spin transport (in devices based on spintronics) and molecular transport (relevant in biological and chemical sensors, membrane or molecular separations and nanofluidity). Examples of nanostructures and nanostructured systems created by self-assembly include arrays of quantum dots, nano- and microelectro-mechanical systems (NEMS and MEMS), adsorbents and catalytic nanoporous materials, nanocrystals and biomimetic materials. The role of mathematics and computer science affects the understanding, control, design and optimisation of these nanostructured materials and functional systems based on nanoscale structures.

In every field of science, there is the need to estimate the parameters of the system to find the extremes of the function object to study, such as a minimum of energy and/or a maximum of entropy and to design and control such systems; mathematical models are so a guide to experimentation and for predicting right characteristics.

For getting a quantitative understanding of nanosystems at a fundamental level, it is essential to have the possibility to calculate the ground state of the system; it can be formulated as a minimisation problem of energy, where the solution gives the configuration of the system particles at lower energy. Complex nanosystems can have millions (or billions) of particles, leading to huge optimisation problems, characterised by a very high number of local minima, with energy levels close to the ground state. Although optimisation techniques provide mainly the answers to questions of predictability, creation of other statistical methods is very important to develop a final statement of the confidence in the response, recognising all possible sources of errors in the process. Modular optimisation, statistical algorithms and software must be developed in a context that provides tools for scientific understanding.

## 1.5 Spintronics, Information Technologies and Nanotechnology

Spintronics is a new branch of electronics in which the spin of the electron (rather than, or in addition to electric charge) is the focus for the storage and transport of information. The spintronic devices have the potential to replace and complete various conventional electronic devices with improved performance. In a broad sense, spintronics includes also new areas, such as spin-based quantum computing and communication. To determine the flexibility of spintronic devices and, more generally, of various applications of spin-polarised transport (like solid-state quantum computing), it is necessary to resolve problems such as the creation and identification of spin-polarised carriers, the maintenance of their spin polarisation and spin coherence, so as how to destroy them.

Dilute magnetic semiconductors attracted considerable attention because of the use of electronic spin, in addition to the charge, for creating new classes of spintronic semiconductor devices with unprecedented functionality. Suggested applications include spin field-effect transistors, semiconductor-based spin valves, non-volatile memory semiconductor chip and spin qubits, to be used as basic building blocks of quantum computing. At nanoscale, spintronic devices give important applications, in relation to magnetic properties of quantum dots and evolution of ferromagnetic properties with the size.

Even the relaxation time of the electronic spin of conduction determines the quality of spintronic devices; as long as the conduction electrons remain in a determined spin state, as long they can store and carry the information. Electrons and nuclear spins are quantum bits (qubits) of quantum computers; the precise manipulation of the dynamics of these spins, in particular to rotate individual spins and to entangle two spins, is one of the most important goals. Electrons can be spin polarised by the light and transfer the polarisation to the nuclei by the hyperfine coupling. In the past it has been proposed and in some cases verified, that a nuclear polarisation is possible saturating the spin of the conduction electrons through a radio frequency field (Overhauser effect), or purely electronically generating hot charges (Feher effect).

Added to electronics based on charge, spintronics brings to an increase of data processing speed, to a decrease of the electric power consumption, compared to traditional semiconductor devices. To successfully incorporate spins in semiconductor existing technology, there are technical problems to be solved, such as efficient injections, transport, control, manipulation and detection of spin polarisation and spin-polarised currents, so as optimisation of spin lifetime, spin coherence detection in nanoscale structures, spin-polarised charge transport for relevant distances and hetero-interfaces, manipulation of contemporary electrons and nuclear spins in sufficiently short times, development of optical methods for spin injection, detection and manipulation.

Considering the superposition of electronics, magnetics and photonics, the idea is to create and improve new multifunctional spin-based devices: spin-FETs (field-effect transistors), spin-LEDs (light emitting diodes) and spin-RTDs (resonant tunnelling devices). The success depends on a deep understanding of the fundamental spin interactions in solid state materials. Understanding and monitoring the spin degree of freedom in semiconductors, semiconductor hetero-structures, ferromagnets and efficiency of spin-based electronics will grow significantly.

### 1.5.1 Spin Decoherence in Electronic Materials

The conduction electrons lose the memory of their spin orientation through collisions with phonons, electrons and other impurities. The spin-orbit interaction provides the necessary spin-dependent potential. This interaction is a relativistic effect that can have various sources in electronic materials, the most important being the electrons–impurities and electrons–ions interactions. This mechanism of spin relaxation in metals and semiconductors is called *Elliott–Yafet mechanism*. The electron spins are a very promising modality for the storage of information.

At low temperatures ($T \leq 20$ K), the spin relaxation is dominated by scattering with impurities and is temperature-independent; for higher temperatures, the electrons lose the spin coherence colliding with phonons. In metals, the spin relaxation times are typically of the order of nanoseconds. A way for growing these relaxation times at low temperatures is to produce very pure samples. The most interesting region is that of room temperatures; in this case, phonons and impurities are the limiting factor. Other possibilities for controlling the spin relaxation include pressure applications, change of dimensionality of the system and doping. Any effect that changes the topology of the Fermi surface will have a significant effect on the relaxation of spin (to be experimentally verified still good).

### 1.5.2 Transport of Polarised Spin in Hybrid Semiconductor Structures

The presence of spin-polarised carriers gives rise to both modified charge transport and intrinsic spin transport, not present in the unpolarised case. Each of these aspects provides information about the degree of spin polarisation that can be utilised in spintronics. The study of semiconductor/superconductor (Sm/S) hybrid structures has several important ramifications. Already in the context of spin-unpolarised transport, it has been demonstrated that this configuration can be used to examine the interfacial transparency, that for a Sm or normal metal is typically limited by a native Schottky barrier. In the presence of spin-polarised carriers, Sm or S structures can also serve to quantify the degree of spin polarisation of a semiconductor and probe both potential and spin-flip interfacial scattering. For a deep understanding of such sensitivity to spin polarisation and different types of interfacial scattering, it is important to consider the process of Andreev reflection that governs the low bias transport. The study of new Sm/S (hybrid structures) can be used to investigate the flexibility of new spintronic devices. It is important to understand the influence of the interfaces between different materials. Considering that the mean free path of the charges overcomes the size of the system, the scattering with interfaces plays a dominant role.

### 1.5.3 Spin-Based Solid State Quantum Computing

It is important to have short ways for moving the information in relation to the spin. Such transfer may be realised by means of close interactions, such as those among nuclear spins, or using mobile objects as the conduction electrons in the semiconductors. The second approach gives more freedom to manipulate the system, but it is also more susceptible to relaxation caused by the transport. One of the first proposals for using electron spin in quantum computing suggested the confinement of electrons in quantum dots, with spins of trapped electrons serving as qubits. With an electron for each quantum dot, each qubit can be readily identified. Individual electron spins can be easily manipulated by a local pulse magnetic field. The controlled exchange interaction among electrons in neighbouring quantum dots can produce entanglement between electron spins. The electron spins relax faster (nanoseconds or microseconds) than nuclear spins (minutes or hours). In the spirit of converting spin information in transport properties, an approach would seem to inject two electron flows in two coupled quantum dots, with one flow totally polarised. Another interesting property combines the extremely long coherence time of nuclear spins with the great industrial silicon experience for producing a scalable quantum computing. The donor nuclear spins are used as qubits in this scheme and also the donor electrons play an important role. Electrons are used to fix the nuclear resonance frequency for one-qubit operations and for transferring information among donor nuclear spins through electronic exchange and hyperfine interaction, crucial for two-qubit operations.

### 1.5.4 Spin Entanglement in Solids

The spin entanglement is an essential ingredient for spin-based quantum computers, for quantum communication, quantum cryptography and other applications. Theoretically, the spin entanglement of two electrons can be measured 'interfering' on an ordinary electron beam or considering the co-tunnelling; it is important to look at the quantum properties of coupled qubits, because it helps the experimental characterisation of quantum computing systems. In quantum computation occurring on these chips, at least two qubits interact in a quantum mechanical way, the 'two-qubit processes'. Other proposals distinguish singlet and triplet states via identification of their energy difference.

The common theme affects the measure of electron transport properties for deducing spin information by transport. The direct spin measurement seems not possible with a superconducting quantum interference device (SQUID) technology [to indicate a device based on the direct current Josephson effect (dc SQUID) or alternate current (RF SQUID), used in many applications], it is slow and not yet sensitive enough for the purposes of quantum computation. If the entangled electrons can be extracted by the superconductor, they can be separated using Stern-Gerlach type techniques and propagate them in separate channels.

### 1.5.5 Optical and Electronic Control of Nuclear Spin Polarisation

The discovery of J. M. Kikkawa and D. D. Awschalom of the optically induced nuclear spin polarisation in GaAs gave rise to the research for new ways to control nuclear spin coherent dynamics. A sample of GaAs kept at 5 K was placed in a magnetic field of about 5 T. Short laser pulses (around 100 fs) of circularly polarised light tuned to the frequency shift of GaAs band (1.5 eV) were sent to the sample, perpendicularly to the applied magnetic field, for creating a population of non-equilibrium spin-polarised conduction electrons.

In the experiments of standard nuclear magnetic resonance, the nuclear spins can be 'adjusted' by applying a microwave radiation of the frequency of spin rotation. The hyperfine interaction is therefore the request oscillating interaction and would allow to invert the nuclear spins. Feher, for the first time, proposed some modalities, such as hot electrons transport, electronic motion in an electric field gradient or in an orthogonal magnetic field, an electron injection whose g-factor differed from that of the electrons in the sample. All these methods are based on the fact that the spin equilibrium proceeds more slowly with respect to the moment equilibrium. The Feher effect shows how to manipulate nuclear spins in a purely electronic way (without the need of optical fields or radio frequency) and can be of great interest in efforts to integrate standard electronics with quantum information processing.

### 1.5.6 Physics of Computation

Computational physics searches to understand the possibilities for processing information by devices with atomic-size elements. Charles Bennett suggested a model of reversible Turing machine, noting a similarity of this model with DNA and ribonucleic acid (RNA). This machine has a direct generalisation on quantum systems (Fig. 1.13).

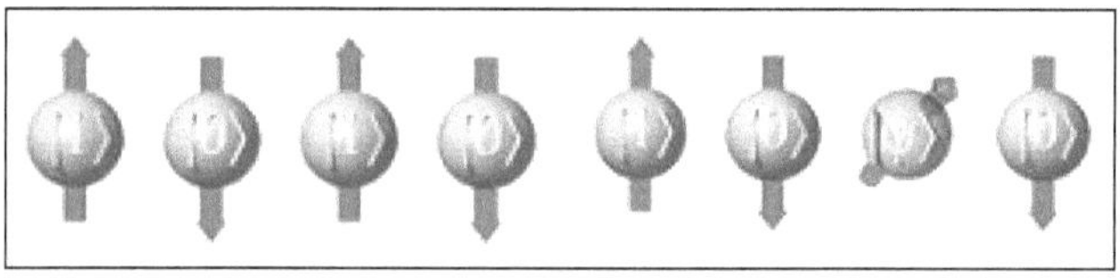

**Figure 1.13** Simulation of a quantum spin chain.

Qubits are the two basic states of a quantum system, $|0\rangle$ and $|1\rangle$. For a system with $N$ particles, there are $2^N$ basic states, in agreement with the principles of quantum mechanics. The classical bits can be represented in the classical model as a discrete set of $2^N$ elements within a space with $2N$ continuous parameters; in the quantum model, the $2^N$ elements 'live' within a $2^N$ space. The difference ($2^N - 2N$) is a rough estimate of 'non-classicity' and grows rapidly with $N$. This complexity consideration is relevant to the presented theme, because the nanoscale domain describes aggregates with more than one quantum systems, but normally is still not enough big for using statistical laws. Quantum information theory provides a convenient language for describing systems with a not overly large number of elements, with an appropriate level of abstraction.

### 1.5.7 Quantum Signal Propagation in Nanosystems

If we consider, for example, the properties of light absorption by certain microorganisms, the importance of quantum effects in this case is quite impossible to disprove. About the effective transfer of photonic energy absorbed by a nanosystem to several elements, in biological systems the efficiency can be of the order of 99%; the uncertainty principle would seem apparently to hinder the optimal transfer. The bio-physical processes in these systems are very similar to the effective signal transmission in a nanodevice, taking into account the quantum effects. The uncertainty of positions and trajectories at quantum mechanics level need special considerations. Some classic models are very well tested at macro-level, but it is unclear if they can also serve in nanosystems.

Quantum information science is often referred to quantum systems with a finite number of states. In the more general case, the term *qudit* is often used for a quantum system with $d$ states; for example, a particle with spin $s$ corresponds to $d = 2s + 1$. The *qubit* is the particular case in which $d = 2$ and $s = 1/2$. Another qudit model is a particle in lattice with $d$ positions; two simple examples with $d$ nodes are the ring (circular system) and the line (linear system) (Fig. 1.14).

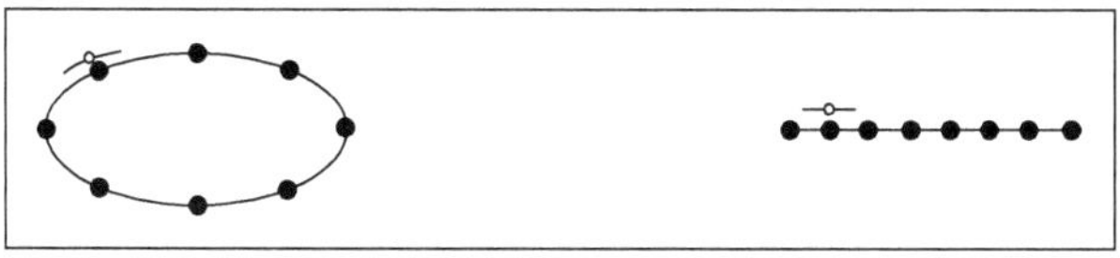

**Figure 1.14** Two simple lattice models.

These chains can be used for quantum communications. In the continuous case, the ideal transmission of a signal might be obtained with a linear law of dispersion; for the quantum case with discrete lattice there is a similar approach. Denoting a state with the occupation of only $t$-th node as $|l\rangle$, it is possible to use a chain of spin instead of lattice with $d$ states. The chain has $2^d$ basic states, only a $d$-dimensional subspace is used and shows a quite standard correspondence between these lattices and the spin waves in the chains with exchange interaction. If we consider a ring with $d$

nodes, the ideal transfer scheme could be described by a cyclic-shift operator:

$$U : |l\rangle \mapsto |l+1 \mod d\rangle, \quad l=0,1,.....,d-1, \tag{1.1}$$

producing a basis with *d* states. The shift operator *U* (Eq. 1.1) is an important attribute of quantum mechanics and it is used in the Weyl commutation relations. In the continuous case, this formalism has a direct correspondence with the Heisenberg commutation relations. Qubits and qudits are considered 'discrete quantum variables', widely used in quantum information science. The technologies of quantum information may be useful for the creation of devices in nanotechnology, because they provide a compact and universal understanding way of the different processes with quantum systems. There is an analogy with the usual information technology application for describing in symbolic form classical objects and processes.

# Chapter 2

# Vector Analysis

## 2.1 Vectors and Scalars

Vector methods are standard tools for physics. In this chapter, we discuss the properties of vectors and vector fields that occur in classical physics. We will do this in a notation that leads to the formation of abstract linear vector spaces.

A physical quantity that is completely specified, in appropriate units, by a single number (called its 'magnitude') such as volume, mass and temperature is called a *scalar*. Scalar quantities are treated as ordinary real numbers. They obey all the regular rules of algebraic addition, subtraction, multiplication, division and so on.

There are also physical quantities that require a magnitude and a direction for their complete specification. These are called *vectors* if their combination with each other is commutative (that is, the order of addition may be changed without affecting the result). Thus, not all quantities possessing magnitude and direction are vectors. Angular displacement, for example, may be characterised by magnitude and direction, but is not a vector: for the addition of two or more angular displacements is not, in general, commutative.

We shall denote vectors by boldface letters (such as $\boldsymbol{a}$) and use ordinary italic letters (such as $a$) for their magnitudes; in writing,

*Mathematics and Physics for Nanotechnology: Technical Tools and Modelling*
Paolo Di Sia

ISBN 978-981-4800-02-0 (Hardcover), 978-0-429-02775-8 (eBook)
www.panstanford.com

vectors are usually represented by a letter with an arrow above it such as $\vec{a}$, or $\overrightarrow{AB}$.

A vector quantity may be graphically represented by an arrow-tipped line segment. The length of the arrow represents the magnitude of the vector and the direction of the arrow is that of the vector. Alternatively, a vector can be specified by its components (projections along the coordinate axes) and the unit vectors (versors) along the coordinate axes (Fig. 2.1):

$$\boldsymbol{a} = a_1\boldsymbol{e}_1 + a_2\boldsymbol{e}_2 + a_3\boldsymbol{e}_3 = \sum_{i=1}^{3} a_i\boldsymbol{e}_i\,, \tag{2.1}$$

where $\boldsymbol{e}_i$ ($i$ = 1, 2, 3) are versors along the rectangular axes $\boldsymbol{x}_i$ and $x_1 = x$, $x_2 = y$, $x_3 = z$; they are normally written as $\boldsymbol{i}$, $\boldsymbol{j}$, $\boldsymbol{k}$ in general physics textbooks.

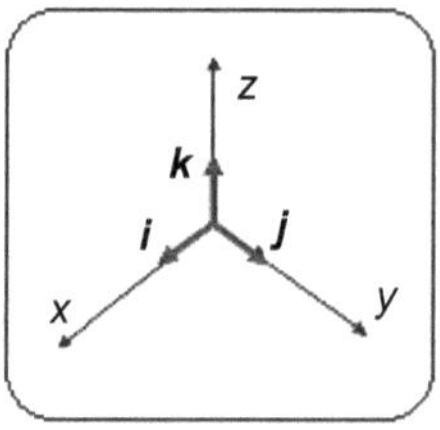


**Figure 2.1** 3-D axes with respective versors.

The component triplet ($a_1$, $a_2$, $a_3$), is also often used as an alternate designation for a vector $\boldsymbol{a}$:

$$\boldsymbol{a} = (a_1, a_2, a_3)\,. \tag{2.2}$$

A vector can be decomposed in components (Fig. 2.2).

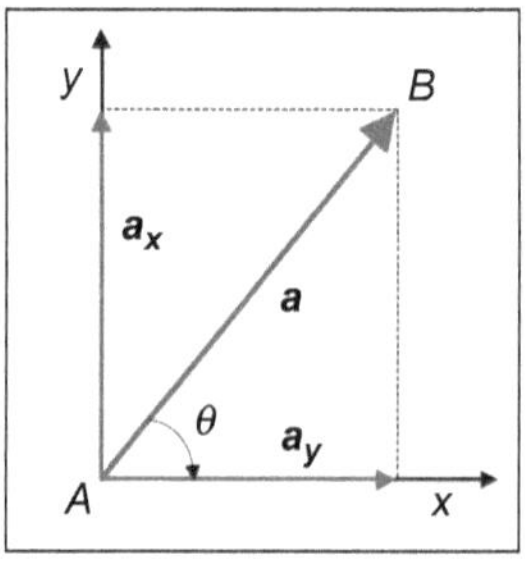


**Figure 2.2** Components of a vector.

The algebraic notation of a vector can be generalised to spaces of dimension greater than three, where an ordered $n$-tuple of real numbers, $(a_1, a_2, \ldots, a_n)$, represents a vector. Even though we cannot construct physical vectors for $n > 3$, we can retain the geometrical language for these $n$-dimensional generalisations.

## 2.2 Direction Angles and Direction Cosines

We can write the generic vector $\boldsymbol{a}$ as a function of cosines of angles formed by the vector with the three Cartesian axes:

$$\boldsymbol{a} = a(\cos\alpha \boldsymbol{e}_1 + \cos\beta \boldsymbol{e}_2 + \cos\gamma \boldsymbol{e}_3). \tag{2.3}$$

In general, the direction cosines of a line are the cosines of the convex angles that the line forms with the Cartesian coordinate axes. The line can be seen in the Cartesian plane (2-D) or in the Euclidean space (3-D) (Fig. 2.3).

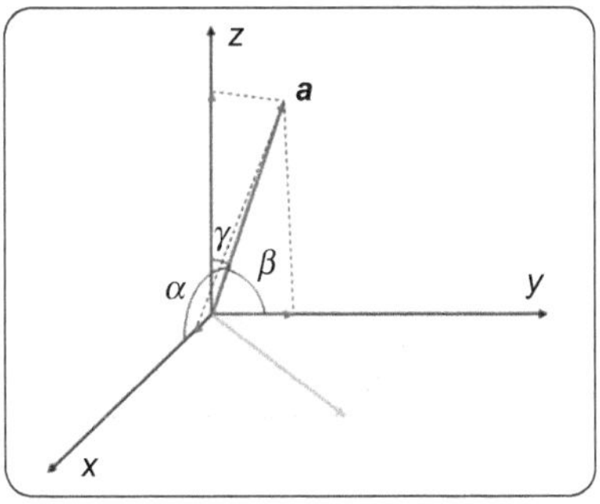


**Figure 2.3** Direction angles of a vector.

## 2.3 Equality of Vectors

Two vectors, say $\boldsymbol{a}$ and $\boldsymbol{b}$, are equal if and only if their respective components are equal:

$$\boldsymbol{a} = \boldsymbol{b} \text{ or } (a_1, a_2, a_3) = (b_1, b_2, b_3). \tag{2.4}$$

This is equivalent to the three equations:

$$a_1 = b_1;\ a_2 = b_2;\ a_3 = b_3. \tag{2.5}$$

Geometrically, equal vectors are parallel and have the same length, but do not necessarily have the same position.

## 2.4 Vector Addition and Subtraction

The addition of two vectors is defined by the equation:

$$\boldsymbol{a} + \boldsymbol{b} = (a_1, a_2, a_3) + (b_1, b_2, b_3) = (a_1 + b_1, a_2 + b_2, a_3 + b_3), \quad (2.6)$$

that is, the sum of two vectors is a vector whose components are sums of the components of the two given vectors.

We can add two non-parallel vectors by a graphical method, also called *parallelogram rule* (Fig. 2.4).

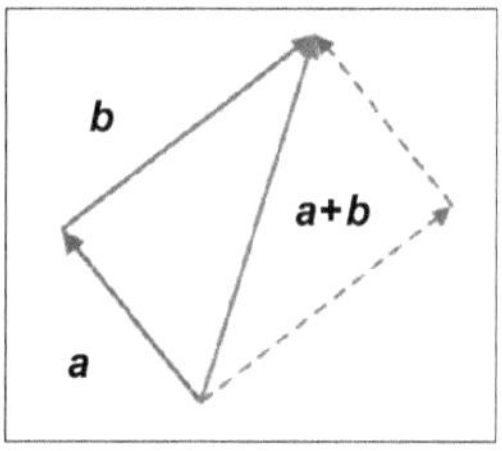


**Figure 2.4** Sum of two non-parallel vectors.

To add vector $\boldsymbol{b}$ to vector $\boldsymbol{a}$, shift $\boldsymbol{b}$ parallel to itself until its tail is at the head of $\boldsymbol{a}$. The vector sum $\boldsymbol{a} + \boldsymbol{b}$ is a vector $\boldsymbol{c}$ drawn from the tail of $\boldsymbol{a}$ to the head of $\boldsymbol{b}$. The order in which the vectors are added does not affect the result.

The vector $-\boldsymbol{a}$ is called the *opposite vector*; its direction is the reverse of that of vector $\boldsymbol{a}$, but both have the same length. Thus, subtraction of vector $\boldsymbol{b}$ from vector $\boldsymbol{a}$ is equivalent to adding $-\boldsymbol{b}$ to $\boldsymbol{a}$:

$$\boldsymbol{a} - \boldsymbol{b} = \boldsymbol{a} + (-\boldsymbol{b}). \quad (2.7)$$

Vector addition has the following properties:

(a) $\boldsymbol{a} + \boldsymbol{b} = \boldsymbol{b} + \boldsymbol{a}$ (commutativity) (2.8)

(b) $(\boldsymbol{a} + \boldsymbol{b}) + \boldsymbol{c} = \boldsymbol{a} + (\boldsymbol{b} + \boldsymbol{c})$ (associativity) (2.9)

(c) $\boldsymbol{a} + \boldsymbol{0} = \boldsymbol{0} + \boldsymbol{a} = \boldsymbol{a}$ (existence of neutral element for addition) (2.10)

(d) $\boldsymbol{a} + (-\boldsymbol{a}) = \boldsymbol{0}$ (existence of opposite element for addition) (2.11)

With the parallelogram rule, we get the 'sum vector' and the 'difference vector' (Fig. 2.5).

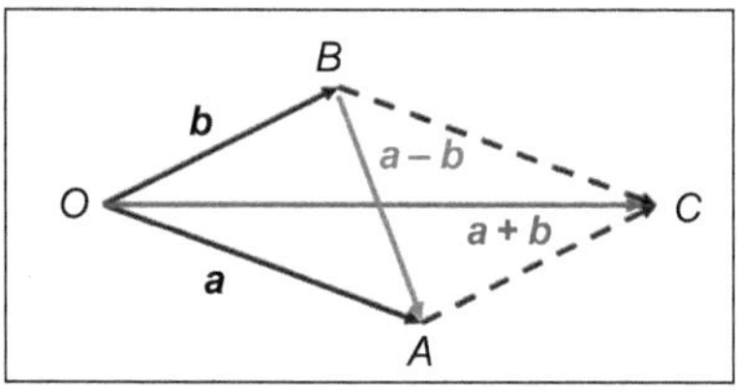


**Figure 2.5** Addition and subtraction of vectors.

## 2.5 Multiplication by a Scalar

If $c$ is a scalar, then:

$$c\,\boldsymbol{a} = (c\,a_1, c\,a_2, c\,a_3). \quad (2.12)$$

Geometrically, the vector '$c\,\boldsymbol{a}$' is parallel to $\boldsymbol{a}$ and is $c$ times the length of $\boldsymbol{a}$. The division by a vector is not defined; expressions such as $k/\boldsymbol{a}$ or $\boldsymbol{b}/\boldsymbol{a}$ are meaningless.

There are several ways of multiplying two vectors, each of which has a special meaning; two types are in particular important in such context: the 'scalar product' and the 'vector product'. The first operation leads to a scalar and the second to a new vector with special features.

## 2.6 Scalar Product

The scalar product (said also *dot product* or *inner product*) of two vectors $\boldsymbol{a}$ and $\boldsymbol{b}$ is a real number defined (in geometrical language) as the product of their magnitude times the cosine of the (smaller) angle between them (Fig. 2.6):

$$\boldsymbol{a} \cdot \boldsymbol{b} = a\,b \cos\theta \; (0 \leq \theta \leq \pi). \quad (2.13)$$

It is normally indicated with a dot '·' between the two vectors.

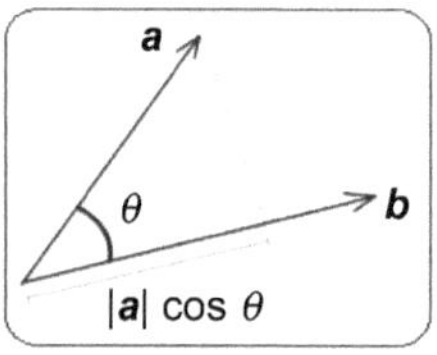


**Figure 2.6** Scalar product of two vectors.

From the definition, we deduct that the scalar product is commutative:

$$\boldsymbol{a} \cdot \boldsymbol{b} = \boldsymbol{b} \cdot \boldsymbol{a}, \tag{2.14}$$

and the product of a vector with itself gives the square of the scalar product of the vector

$$\boldsymbol{a} \cdot \boldsymbol{a} = a^2. \tag{2.15}$$

If $\boldsymbol{a} \cdot \boldsymbol{b} = 0$ and neither $\boldsymbol{a}$ nor $\boldsymbol{b}$ is a null (zero) vector, then $\boldsymbol{a}$ is perpendicular to $\boldsymbol{b}$.

We can get a simple geometric interpretation of the scalar product considering Fig. 2.6:

$(b \cos \theta)\, a$ = projection of $\boldsymbol{b}$ onto $\boldsymbol{a}$ multiplied by the magnitude of $\boldsymbol{a}$;

$(a \cos \theta)\, b$ = projection of $\boldsymbol{a}$ onto $\boldsymbol{b}$ multiplied by the magnitude of $\boldsymbol{b}$.

If only the components of $\boldsymbol{a}$ and $\boldsymbol{b}$ are known, then it would not be practical to calculate $\boldsymbol{a} \cdot \boldsymbol{b}$ from Eq. 2.13. In this case, we can calculate $\boldsymbol{a} \cdot \boldsymbol{b}$ in terms of the components:

$$\boldsymbol{a} \cdot \boldsymbol{b} = (a_1\boldsymbol{e}_1 + a_2\boldsymbol{e}_2 + a_3\boldsymbol{e}_3) \cdot (b_1\boldsymbol{e}_1 + b_2\boldsymbol{e}_2 + b_3\boldsymbol{e}_3), \tag{2.16}$$

the right-hand side has nine terms, all involving the product $\boldsymbol{e}_i \cdot \boldsymbol{e}_j$. We remember that the angle between each pair of unit vectors is 90°, so we can write:

$$\boldsymbol{e}_i \cdot \boldsymbol{e}_j = \delta_{ij}, \quad i, j = 1, 2, 3 \tag{2.17}$$

where $\delta_{ij}$ is the Kronecker delta symbol:

$$\delta_{ij} = \begin{cases} 0, & if \ i \neq j \\ 1, & if \ i = j \end{cases}. \tag{2.18}$$

The calculation of the right part of Eq. 2.16 brings to the result:

$$\boldsymbol{a}\cdot \boldsymbol{b} = a_1b_1 + a_2b_2 + a_3b_3 = \sum_{i=1}^{3} a_i b_i \ . \qquad (2.19)$$

## 2.7 Vector Product

The vector product of two vectors $\boldsymbol{a}$ and $\boldsymbol{b}$ gives a vector and is written as:

$$\boldsymbol{c} = \boldsymbol{a} \times \boldsymbol{b}. \qquad (2.20)$$

It is normally indicated with a cross '×'. The direction of the vector $\boldsymbol{c}$ is determined by a rule also called *right-hand rule* (Fig. 2.7).

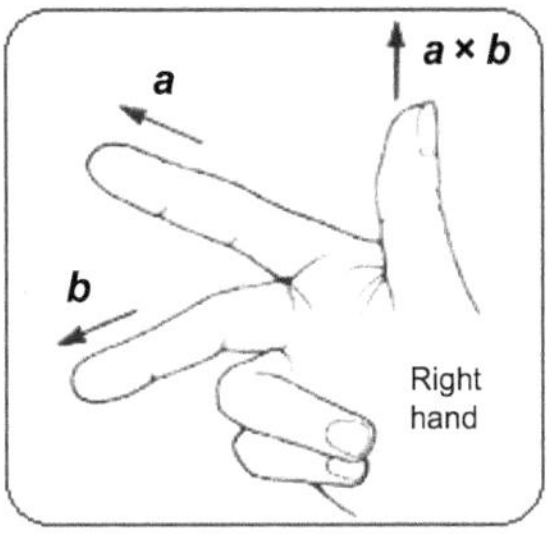


**Figure 2.7** Right-hand rule for vector product.

Vector $\boldsymbol{c}$ is defined to be perpendicular to the plane containing vectors $\boldsymbol{a}$ and $\boldsymbol{b}$, with its magnitude equal to the area of the parallelogram defined by $\boldsymbol{a}$ and $\boldsymbol{b}$. The direction of $\boldsymbol{c}$ is along the thumb of the right hand when the fingers rotate from $\boldsymbol{a}$ to $\boldsymbol{b}$ (angle of rotation less than 180°). The modulus (intensity) of $\boldsymbol{c}$ is, therefore,

$$\mathrm{c} = \mathrm{ab}\sin\theta, \quad (0 \le \theta \le \pi). \qquad (2.21)$$

From the definition of the vector product and following the right-hand rule, we see immediately that

$$\boldsymbol{a} \times \boldsymbol{b} = -\boldsymbol{b} \times \boldsymbol{a}. \qquad (2.22)$$

Hence, the vector product is not commutative. If $\boldsymbol{a}$ and $\boldsymbol{b}$ are parallel, then it follows that the product is zero and also if the angle is zero.

The product can be written as an easily remembered determinant of third order:

$$\boldsymbol{a}\times\boldsymbol{b} = \begin{vmatrix} \boldsymbol{i} & \boldsymbol{j} & \boldsymbol{k} \\ a_x & a_y & a_z \\ b_x & b_y & b_z \end{vmatrix}. \qquad (2.23)$$

## 2.8 Triple Scalar Product

We briefly discuss in this paragraph the scalar $\boldsymbol{a}\cdot(\boldsymbol{b}\times\boldsymbol{c})$. This scalar represents the volume of the parallelepiped formed by the coterminous sides ***a***, ***b***, ***c***, since:

$$\boldsymbol{a}\cdot(\boldsymbol{b}\times\boldsymbol{c}) = abc\sin\theta\cos\alpha = hS, \qquad (2.24)$$

where $S$ being the area of the parallelogram with sides ***b*** and ***c***, and $h$ is the height of the parallelogram (Fig. 2.8).

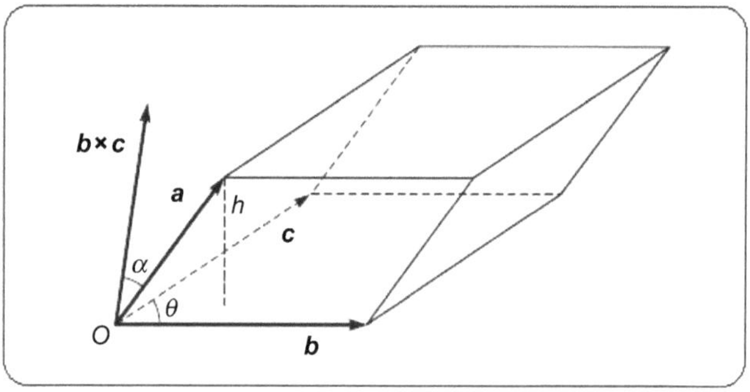


**Figure 2.8** Triple scalar product of vectors ***a***, ***b***, and ***c***.

Considering Eq. 2.23, we can write:

$$\boldsymbol{a}\cdot(\boldsymbol{b}\times\boldsymbol{c}) = \begin{vmatrix} a_x & a_y & a_z \\ b_x & b_y & b_z \\ c_x & c_y & c_z \end{vmatrix}. \qquad (2.25)$$

The exchange of two rows (or two columns) changes the sign of the determinant, but does not change its absolute value. Using this property, we find that the dot and the cross may be interchanged in the triple scalar product:

$$\boldsymbol{a}\cdot(\boldsymbol{b}\times\boldsymbol{c}) = (\boldsymbol{a}\times\boldsymbol{b})\cdot\boldsymbol{c}. \qquad (2.26)$$

In fact, as long as the three vectors appear in cyclic order, then the dot and cross may be inserted between any pairs:

$$\boldsymbol{a} \cdot (\boldsymbol{b} \times \boldsymbol{c}) = \boldsymbol{b} \cdot (\boldsymbol{c} \times \boldsymbol{a}) = \boldsymbol{c} \cdot (\boldsymbol{a} \times \boldsymbol{b}). \tag{2.27}$$

It should be noted that the scalar resulting from the triple scalar product changes sign on an inversion of coordinates. For this reason, the triple scalar product is sometimes called a *pseudoscalar*.

## 2.9 Triple Vector Product

The triple product $\boldsymbol{a} \times (\boldsymbol{b} \times \boldsymbol{c})$ is a vector, since it is the vector product of two vectors: $\boldsymbol{a}$ and $(\boldsymbol{b} \times \boldsymbol{c})$. This vector is perpendicular to $(\boldsymbol{b} \times \boldsymbol{c})$ and it lies in the plane of $\boldsymbol{b}$ and $\boldsymbol{c}$. It holds:

$$\boldsymbol{a} \times (\boldsymbol{b} \times \boldsymbol{c}) = \boldsymbol{b}(\boldsymbol{a} \cdot \boldsymbol{c}) - \boldsymbol{c}(\boldsymbol{a} \cdot \boldsymbol{b}). \tag{2.28}$$

## 2.10 Linear Vector Space *V*

A set $V$ is a *linear vector space* if:

(a) among the elements of $V$, it is defined an operation of internal composition, said 'sum', such that:

$$\forall a \in V, \quad \forall b \in V \rightarrow a + b \in V. \tag{2.29}$$

We say that $V$ is 'closed' with respect to the sum. The following properties hold:

(i) $\forall a,b \in V \rightarrow a + b = b + a$ (commutative property) (2.30)

(ii) $\forall a,b,c \in V \rightarrow (a + b) + c = a + (b + c)$ (associative property) (2.31)

(iii) $\exists ! 0 \in V$: $\forall a \in V \rightarrow a + 0 = 0 + a = a$ (existence and unicity of the neutral element, indicated with 0) (2.32)

(iv) $\forall a \in V \quad \exists !(-a) \in V$: $\quad a + (-a) = 0$ (existence and unicity of the opposite element $a \in V$, indicated with $-a$) (2.33)

(b) it is defined an operation said 'multiplication' among real numbers and elements of $V$, such that:

$$\forall k \in \Re, \quad \forall a \in V \rightarrow ka \in V. \tag{2.34}$$

We say that $V$ is closed with respect to this multiplication; the following properties hold:

$$\begin{array}{llll} \text{(i)} & & \forall a \in V & 1a = a; \\ \text{(ii)} & \forall h, k \in \Re & \forall a \in V & h(ka) = (hk)a; \\ \text{(iii)} & \forall h, k \in \Re & \forall a \in V & (h + k)a = ha + ka; \\ \text{(iv)} & \forall k \in \Re & \forall a, b \in V & k(a + b) = ka + kb. \end{array} \quad (2.35)$$

A (*vector*) *subspace* is a subset $S$ of a vector space, i.e., it is closed with respect to the operations of sum among its elements and of multiplication above defined. For these operations, the properties of vector space hold.

In $\Re^n$, we say 'vector' an ordered $n$-tuple of real numbers: $\mathbf{x} = (x_1, x_2,\ldots, x_n)$ . $\Re^n$ is the set of ordered $n$-tuples of real numbers, i.e., the Cartesian product of $\Re$ by itself $n$ times:

$$\Re^n = \{\mathbf{x} = (x_1, x_2,\ldots, x_n): \quad x_i \in \Re \quad i = 1,2,\ldots, n\}. \quad (2.36)$$

Then $\mathbf{x}$ is an element, a point, of $\Re^n$. Also in $\Re^n$ the sum is defined:

$$\forall \mathbf{x} = (x_1, x_2,\ldots, x_n) \in \Re^n; \forall \mathbf{y} = (y_1, y_2,\ldots, y_n) \in \Re^n \rightarrow$$

$$\mathbf{z} = \mathbf{x} + \mathbf{y} = (x_1 + y_1, x_2 + y_{2,\ldots,} x_n + y_n) \in \Re^n. \quad (2.37)$$

For this operation, commutative and associative properties hold, so as the existence of neutral element [it is the null vector $\mathbf{0} = (0, 0,,\ldots,0)$]; the opposite of $\mathbf{x} = (x_1, x_2,\ldots,x_n)$ is $-\mathbf{x} = (-x_1, -x_2,\ldots,-x_n)$ .

## Exercises

2.1 Verify that the null vector $\boldsymbol{0} = (0, 0,,\ldots,0) \in \Re^n$ is orthogonal to any vector $\mathbf{x} \in \Re^n$ .

2.2 Find for what values of the real parameter $a$, the vector $\mathbf{v} = (a, a - 1, 2)$ is orthogonal to vector $\mathbf{w} = (1, 2, 3)$. [$a = -4/3$]

2.3 Determine the vector, that is parallel to $\mathbf{v} = (1, - 1, -2)$ and with magnitude = 1. [$\mathbf{u} = \sqrt{1/6}\mathbf{v}$]

2.4 Calculate the direction cosines of vector $\boldsymbol{a} = 3\boldsymbol{i} + 4\boldsymbol{j}$ .
[$\cos\alpha = 3/5$; $\cos\beta = 4/5$]

2.5 A particle makes three consecutive shifts:

$\mathbf{s}_1 = (1.5\boldsymbol{i} + 3.0\boldsymbol{j} - 1.2\boldsymbol{k})$ cm;

$\boldsymbol{s}_2 = (2.3\boldsymbol{i} - 1.4\boldsymbol{j} - 3.6\boldsymbol{k})$ cm;

$\boldsymbol{s}_3 = (-1.3\boldsymbol{i} + 1.5\boldsymbol{j})$ cm.

Calculate magnitude and direction of resultant displacement.

[$R = 6.2$ cm; $\theta = \text{ar cos}\,(R_z/R)$; $\phi = \text{arc tan}\,(R_y/R_x)$]

2.6 Given the two vectors: $\boldsymbol{a} = 2\boldsymbol{i} - \boldsymbol{j} - 3\boldsymbol{k}$; $\boldsymbol{b} = -2\boldsymbol{i} + 2\boldsymbol{j} - \boldsymbol{k}$; calculate $(\boldsymbol{a} - 3\boldsymbol{b}) \times (-2\boldsymbol{a} + \boldsymbol{b})$. [$\boldsymbol{c} = -35\boldsymbol{i} - 40\boldsymbol{j} - 10\boldsymbol{k}$]

2.7 Given the two vectors: $\boldsymbol{a} = \boldsymbol{i} - \boldsymbol{k}$; $\boldsymbol{b} = 2\boldsymbol{i} + 4\boldsymbol{j} + 3\boldsymbol{k}$; calculate

(a) their scalar and vector product;

(b) the angle between the two vectors.

[$\boldsymbol{a} \cdot \boldsymbol{b} = -1$; $\boldsymbol{a} \times \boldsymbol{b} = 4\boldsymbol{i} - 5\boldsymbol{j} + 4\boldsymbol{k}$; $\alpha = \text{ar cos}\,(-1/\sqrt{58})$]

2.8 Determine the angle $\theta$ between the two vectors $\boldsymbol{a} = k(6\boldsymbol{i} - 3\boldsymbol{j})$ and $\boldsymbol{b} = \boldsymbol{a} + \boldsymbol{i}$ ($k$ is a real number). Find the value of $\theta$ if $k \to \infty$.

[$k \to \infty \Rightarrow \cos\theta = 1$]

# Chapter 3

# Vector Differentiation

## 3.1 Introduction

A vector may be a function of one or more scalars and vectors. There are, for example, many important vectors in mechanics that are functions of time and position variables. In physics, the concept of field is often used for representing a physical quantity that is a function of position in a given region. Temperature is a *scalar field*, because its value depends upon location: each point $P = (x, y, z)$ is associated with a temperature $T(x, y, z)$. The function $T(x, y, z)$ is a scalar field, whose value is a real number depending only on the point in space, but not on the particular choice of the coordinate system.

A *vector field*, on the other hand, associates a vector to each point (that is, three numbers at each point are associated), such as the wind velocity or the strength of the electric or magnetic field. When described in a rotated system, for example, the three components of the vector associated with one and the same point will change in numerical value. Physically and geometrically important concepts in connection with scalar and vector fields are the 'gradient', 'divergence', 'curl' and the corresponding 'integral theorems'.

The basic concepts of calculus, such as continuity and differentiability, can be naturally extended to vector calculus.

*Mathematics and Physics for Nanotechnology: Technical Tools and Modelling*
Paolo Di Sia

ISBN 978-981-4800-02-0 (Hardcover), 978-0-429-02775-8 (eBook)
www.panstanford.com

Let us consider a vector $\boldsymbol{a}$, whose components are functions of a single variable $u$. If, for example, the vector $\boldsymbol{a}$ represents a position or velocity, then the parameter $u$ is usually the time $t$, but it can be any quantity that determines the components of $\boldsymbol{a}$. If we introduce a Cartesian coordinate system, the vector function $\boldsymbol{a}(u)$ may be written as:

$$\boldsymbol{a}\,(u) = a_1\,(u)\boldsymbol{i} + a_2\,(u)\boldsymbol{j} + a_3\,(u)\boldsymbol{k}, \tag{3.1}$$

$\boldsymbol{a}(u)$ is said to be continuous at $u = u_0$ if it is defined in some neighbourhood of $u_0$ and

$$\lim_{u \to u_0} a(u) = a(u_0). \tag{3.2}$$

Note that $\boldsymbol{a}(u)$ is *continuous* at $u_0$ if and only if its three components are continuous at $u_0$.

$\boldsymbol{a}(u)$ is said to be *differentiable* at a point $u$ if the limit:

$$\boldsymbol{a}'(u_0) = \lim_{h \to 0} \frac{a(u_0 + h) - a(u_0)}{h} \tag{3.3}$$

exists. The vector $\boldsymbol{a}'(u)$ is called the *derivative* of $\boldsymbol{a}(u)$. To differentiate a vector function, we differentiate each component separately:

$$a'\,(u) = a'_1\,(u)\boldsymbol{i} + a'_2\,(u)\boldsymbol{j} + a'_3\,(u)\boldsymbol{k}. \tag{3.4}$$

Note that the unit coordinate vectors are fixed in space. Higher derivatives of $\boldsymbol{a}(u)$ can be similarly defined.

If $\boldsymbol{a}$ is a vector depending on more than one scalar variable, say $u$, $v$, then we write $\boldsymbol{a} = \boldsymbol{a}(u, v)$. Then:

$$d\boldsymbol{a} = \frac{\partial \boldsymbol{a}}{\partial u} du + \frac{\partial \boldsymbol{a}}{\partial v} dv \tag{3.5}$$

is the *differential* of $\boldsymbol{a}$ and

$$\frac{\partial \boldsymbol{a}}{\partial u} = \lim_{h \to 0} \frac{\boldsymbol{a}(u+h, v) - \boldsymbol{a}(u, v)}{h} \quad \text{and} \tag{3.6}$$

$$\frac{\partial \boldsymbol{a}}{\partial v} = \lim_{h \to 0} \frac{\boldsymbol{a}(u, v+h) - \boldsymbol{a}(u, v)}{h} \tag{3.7}$$

are the two prime partial derivatives.

Derivatives of products obey rules similar to those for scalar functions. However, when cross products are involved, the order may be important.

## 3.2 The Gradient Operator

Given a scalar field $S(x, y, z)$, let us consider a generic point $P$ of the domain of $S$; if we do an infinitesimal displacement from $P$ to $P'$, identified with the infinitesimal vector:

$$d\boldsymbol{l} = dx\boldsymbol{i} + dy\boldsymbol{j} + dz\boldsymbol{k} \tag{3.8}$$

$S$ varies accordingly, except for infinitesimals of higher order of the quantity:

$$dS = \frac{\partial S}{\partial x}dx + \frac{\partial S}{\partial y}dy + \frac{\partial S}{\partial z}dz\,. \tag{3.9}$$

$dS$ is a scalar quantity and $(dx, dy, dz)$ are the components of a vector. Therefore, if also:

$$\left(\frac{\partial S}{\partial x}, \frac{\partial S}{\partial y}, \frac{\partial S}{\partial z}\right) \tag{3.10}$$

are considered as components of a vector, we can write the scalar product as:

$$dS = grad\,S \cdot d\boldsymbol{l}, \tag{3.11}$$

where with *grad S* we mean the vector:

$$grad\,S = \frac{\partial S}{\partial x}\boldsymbol{i} + \frac{\partial S}{\partial y}\boldsymbol{j} + \frac{\partial S}{\partial z}\boldsymbol{k}\,. \tag{3.12}$$

As a geometric interpretation of *grad S*, we note that $S(x_1, x_2, x_3) = c$, where $c$ is a constant, represents a surface. Let $\boldsymbol{l} = x\boldsymbol{i} + y\boldsymbol{j} + z\boldsymbol{k}$ be the position vector to a point $P = (x_1, x_2, x_3)$ on the surface. If we move along the surface to a nearby point $Q$, then $d\boldsymbol{l} = \boldsymbol{d}xi + dy\boldsymbol{j} + dz\boldsymbol{k}$ lies in the tangent plane to the surface at $P$, but as long as we move along the surface, $S$ has a constant value and therefore, $dS = 0$ (Fig. 3.1). Consequently from Eq. 3.9

$$d\boldsymbol{l}\cdot\nabla S = 0. \tag{3.13}$$

Equation 3.13 states that *grad S* is perpendicular to ***dl*** and therefore to the surface (Fig. 3.2).

The *gradient* identifies the direction of maximum variation of *S*. The change from the scalar *S* to vector ***grad S*** is generally always possible. It is not said instead that, given a vector field, there exists always a scalar function of which that field is the gradient. If this happens, it is said that the vector field is *conservative*.

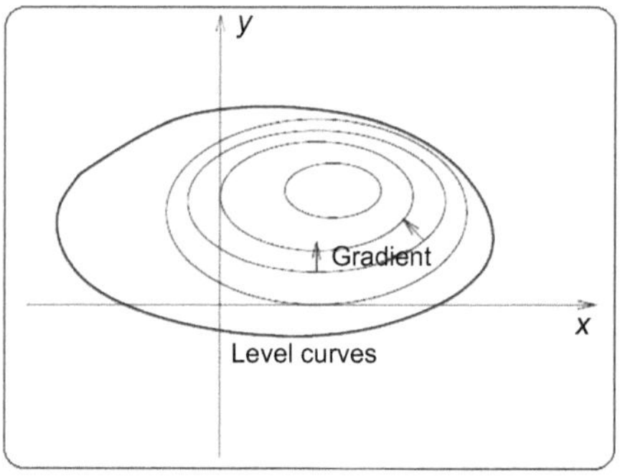


**Figure 3.1** Level curves and gradient (2-D).

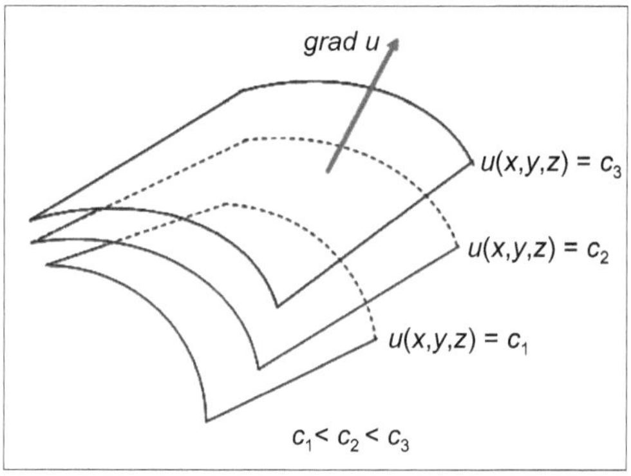


**Figure 3.2** Orthogonality of gradient with respect to the surface (3-D).

It is possible also to consider the symbolic vector:

$$\nabla = \frac{\partial}{\partial x}\boldsymbol{i} + \frac{\partial}{\partial y}\boldsymbol{j} + \frac{\partial}{\partial z}\boldsymbol{k} \tag{3.14}$$

said 'del', or 'nabla'. Acting on a scalar, we find the common definition of gradient:

$$\nabla S = \frac{\partial S}{\partial x}\boldsymbol{i} + \frac{\partial S}{\partial y}\boldsymbol{j} + \frac{\partial S}{\partial z}\boldsymbol{k}. \tag{3.15}$$

The introduction of the symbolic vector $\nabla$ allows to reduce the differential operations to products of the type 'vectors–vectors' or 'vectors–scalars', with the attention that the commutative property does not hold. In this way, we obtain compactness in the calculation and formal elegance.

## 3.3 Directional Derivative

The directional derivative is a tool that generalises the concept of partial derivative of a function of several variables, by extending it to any direction identified by a vector. In differential geometry, the directional derivative is generalised to a differentiable manifold through the concept of 'covariant derivative'.

In $\Re^n$, the directional derivative of a scalar function $f(\mathbf{x}) = f(x_1, x_2,\ldots,x_n)$ along a unit vector $\boldsymbol{u} = (u_1, u_2,\ldots,u_n)$ is defined by the limit:

$$D_{\boldsymbol{u}} f(\boldsymbol{x}) = \lim_{h\to 0} \frac{f(\boldsymbol{x}+h\boldsymbol{u})-f(\boldsymbol{x})}{h}. \tag{3.16}$$

At each point $\mathbf{x}$, the directional derivative $D_{\boldsymbol{u}}f(\mathbf{x})$ represents the change of $f$ along $\boldsymbol{u}$. If the function $f$ is differentiable in $\mathbf{x}$, then the directional derivative exists along every unit vector and it holds:

$$D_{\boldsymbol{u}}f(\mathbf{x}) = \nabla f(\mathbf{x})\cdot\boldsymbol{u}, \tag{3.17}$$

i.e., scalar product of the gradient is $f$ times the unit vector $\boldsymbol{u}$.

The calculation of a directional derivative is, therefore, possible in two ways:

(a) through definition, but this is usually a long operation;

(b) through partial derivatives (Eq. 3.17):

$$\frac{\partial f}{\partial \boldsymbol{u}}(x_0, y_0) = \nabla f(x_0, y_0)\cdot\boldsymbol{u} = u_1\frac{\partial f}{\partial x}(x_0, y_0) + u_2\frac{\partial f}{\partial y}(x_0, y_0). \tag{3.18}$$

***Example* 3.1**: For the given function

$$f(x,y) = \frac{x}{x^2+y^2},$$

determine ***grad f*** in $P_0 = (1,2)$ and the directional derivative of $f$ in $P_0$ along the unit vector

$$\boldsymbol{u} = -\frac{\sqrt{2}}{2}\boldsymbol{i} + \frac{\sqrt{2}}{2}\boldsymbol{j}.$$

***Solution***: The calculation of prime derivatives gives:

$$f'_x = \frac{-x^2+y^2}{(x^2+y^2)^2} \rightarrow f'_x(P_0) = \frac{3}{25}$$

$$f'_y = \frac{-2xy}{(x^2+y^2)^2} \rightarrow f'_y(P_0) = -\frac{4}{25}$$

The searched gradient is therefore

$$\nabla f(1,2) = \frac{3}{25}\boldsymbol{i} - \frac{4}{25}\boldsymbol{j}.$$

## 3.4 The Divergence Operator

Vector differential operations on vector fields are more complicated because of vector nature of both the operator and the field on which it operates. As we know, there are two types of products involving two vectors, namely the scalar and vector products; vector differential operations on vector fields can also be separated into two types called the *divergence* and the *curl*.

Given a vector field $\boldsymbol{a}$, we define *divergence* of the field the scalar quantity:

$$div\,\boldsymbol{a} = \frac{\partial a_x}{\partial x} + \frac{\partial a_y}{\partial y} + \frac{\partial a_z}{\partial z}. \tag{3.19}$$

The operation allows to pass from a vector field to a scalar field. A vector field with divergence always equal to zero is known as *solenoidal* (Fig. 3.3).

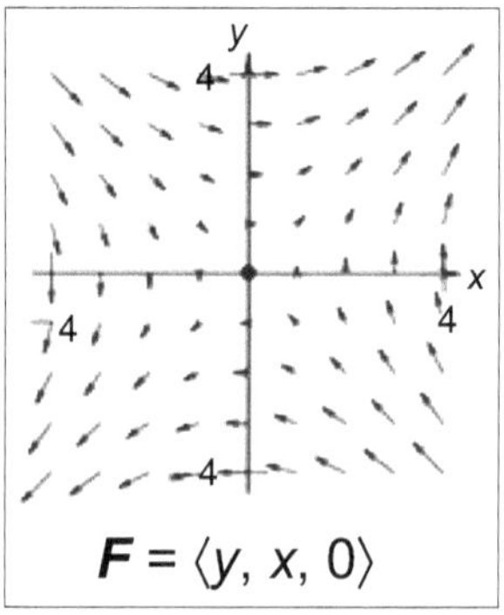


**Figure 3.3** Solenoidal field $\boldsymbol{F} = (y, x, 0)$.

Using the symbolic vector $\nabla$, we can write:

$$\nabla \cdot \boldsymbol{a} = \frac{\partial a_x}{\partial x} + \frac{\partial a_y}{\partial y} + \frac{\partial a_z}{\partial z}. \tag{3.20}$$

The divergence operator is useful in determining whether there is a source or a well in the space areas where vector fields exist. For electromagnetic fields, for example, these sources and wells will turn out to be positive and negative electrical charges. This region could also be located in a river where water would be flowing (Fig. 3.4).

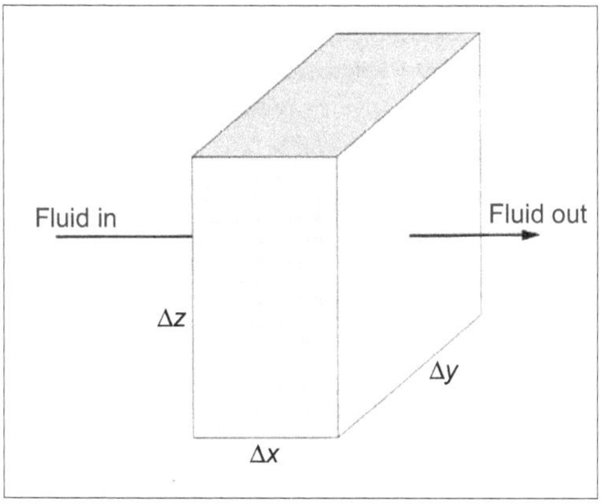


**Figure 3.4** The divergence is linked to the flux.

This could be a very porous box that contained either a drain or faucet that was connected with an invisible hose to the shore, where the fluid could either be absorbed or from which it could be extracted.

## 3.5 The Laplacian Operator

The *Laplacian*, or *Laplace operator*, is obtainable considering the scalar product of operators $\nabla$:

$$\nabla \cdot \nabla = \nabla^2 = \frac{\partial^2}{\partial x^2} + \frac{\partial^2}{\partial y^2} + \frac{\partial^2}{\partial z^2}. \tag{3.21}$$

It is a scalar operator; acting on a scalar, it gives a scalar field and acting on a vector, it gives a vector field.

## 3.6 The Curl Operator

Given a vector field $\boldsymbol{a}$, we define *curl* of $\boldsymbol{a}$ the vector $\boldsymbol{b}$ as:

$$\boldsymbol{b} = curl\,\boldsymbol{a} = \begin{vmatrix} \boldsymbol{i} & \boldsymbol{j} & \boldsymbol{k} \\ \dfrac{\partial}{\partial x} & \dfrac{\partial}{\partial y} & \dfrac{\partial}{\partial z} \\ a_x & a_y & a_z \end{vmatrix}. \tag{3.22}$$

Using the symbolic vector $\nabla$, we have:

$$\nabla \times \boldsymbol{a} = curl\,\boldsymbol{a}. \tag{3.23}$$

This operation allows to pass from a vector field to another vector field. A vector field that has a non-vanishing curl is called a *vortex field* and the curl of the field vector is a measure of the 'vorticity' of the vector field.

A vector field with curl identically equal to zero is known as *irrotational* (Fig. 3.5).

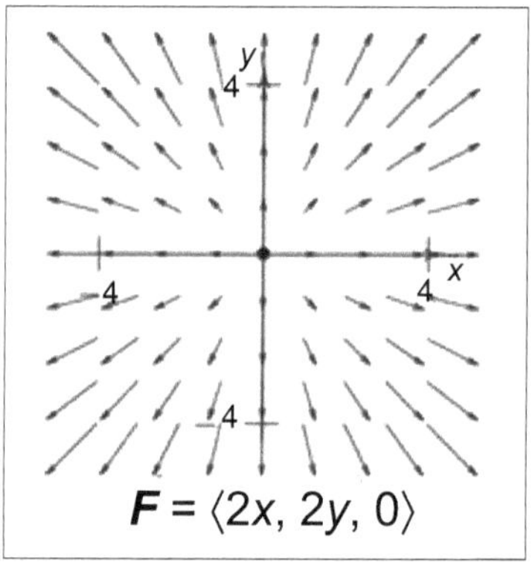


**Figure 3.5** Irrotational field $\boldsymbol{F} = (2x, 2y, 0)$.

## 3.7 Formulas Involving the Nabla Operator

We now list some important formulas involving the vector differential operator 'nabla'. In these formulas, $\boldsymbol{a}$ and $\boldsymbol{b}$ are differentiable vector field functions and $f$ and $g$ are differentiable scalar field functions of position $(x_1, x_2, x_3)$:

(a) $\nabla(fg) = f\nabla g + g\nabla f$; (3.24a)

(b) $\nabla \cdot (f\boldsymbol{a}) = f\nabla \cdot \boldsymbol{a} + \nabla f \cdot \boldsymbol{a}$; (3.24b)

(c) $\nabla \times (f\boldsymbol{a}) = f\nabla \times \boldsymbol{a} + \nabla f \times \boldsymbol{a}$; (3.24c)

(d) $\nabla \times (\nabla f) = 0$; (3.24d)

(e) $\nabla \cdot (\nabla \times \boldsymbol{a}) = 0$; (3.24e)

(f) $\nabla \cdot (\boldsymbol{a} \times \boldsymbol{b}) = (\nabla \times \boldsymbol{a}) \cdot \boldsymbol{b} - (\nabla \times \boldsymbol{b}) \times \boldsymbol{a}$; (3.24f)

(g) $\nabla \times (\boldsymbol{a} \times \boldsymbol{b}) = (\boldsymbol{b} \cdot \nabla)\boldsymbol{a} - \boldsymbol{b}\,(\nabla \cdot \boldsymbol{a}) + \boldsymbol{a}(\nabla \cdot \boldsymbol{b}) - (\boldsymbol{a} \cdot \nabla)\boldsymbol{b}$; (3.24g)

(h) $\nabla \times (\nabla \times \boldsymbol{a}) = \nabla(\nabla \cdot \boldsymbol{a}) - \nabla^2\boldsymbol{a}$; (3.24h)

(i) $\nabla\,(\boldsymbol{a} \cdot \boldsymbol{b}) = \boldsymbol{a} \times (\nabla \times \boldsymbol{b}) + \boldsymbol{b} \times (\nabla \times \boldsymbol{a}) + (\boldsymbol{a} \cdot \nabla)\boldsymbol{b} + (\boldsymbol{b} \cdot \nabla)\boldsymbol{a}$. (3.24i)

## Exercises

3.1 Given the function: $f(x,y) = \dfrac{x}{x^2+y^2}$, calculate the gradient of $f$ in $P_0 = (1, 2)$. $\left[\nabla f(1,2) = \dfrac{3}{25}\boldsymbol{i} - \dfrac{4}{25}\boldsymbol{j}\right]$

3.2 Calculate the curl of the vector field:
$\boldsymbol{F}(x, y, z) = (y + z)\boldsymbol{i} + (x + z)\boldsymbol{j} + (x - y)\boldsymbol{k}$. $[\nabla \times \boldsymbol{F} = -2\boldsymbol{i}]$

3.3 Given: $f(x, y, z) = 2xz^4 - x^2y$, calculate *grad* $f$ and its intensity in $P_1 = (2, -2, 1)$. $\left[\nabla \boldsymbol{f} = 10\boldsymbol{i} - 4\boldsymbol{j} + 16\boldsymbol{k}; |\nabla \boldsymbol{f}| = 2\sqrt{93}\right]$

3.4 Given: $f(x, y, z) = 3x^2 - yz$ and $\boldsymbol{F}(x, y, z) = 3xyz^2\boldsymbol{i} + 2xy^3\boldsymbol{j} - x^2yz\boldsymbol{k}$, calculate (a) $\nabla \cdot \boldsymbol{F}$; (b) $\boldsymbol{F} \cdot \nabla \boldsymbol{f}$ in $P_1 = (1, -1, 1)$. [4, 15]

3.5 Given: $f(x, y, z) = 3x^2 - yz$ and $\boldsymbol{F}(x, y, z) = 3xyz^2\boldsymbol{i} + 2xy^3\boldsymbol{j} - x^2yz\boldsymbol{k}$, calculate: (a) $\nabla \cdot (f\boldsymbol{F})$; (b) $\nabla \cdot (\nabla \boldsymbol{f})$ in $P_1 = (1, -1, 1)$. [1, 6]

3.6 Verify if the vector field $\boldsymbol{F}(x, y, z) = 3y^4z^2\boldsymbol{i} + 4x^3z^2\boldsymbol{j} - 3x^2y^2\boldsymbol{k}$ is solenoidal. [Yes]

3.7 Find the curl of the vector field:
$\boldsymbol{F}(x, y, z) = xz^3\boldsymbol{i} - 2x^2yz\boldsymbol{j} + 2yz^4\boldsymbol{k}$ in $P_1 = (1, -1, 1)$.
$[\boldsymbol{F}(1, -1, 1) = 3\boldsymbol{j} + 4\boldsymbol{k}]$

3.8 Find the Laplacian of the field:
$f(x, y, z) = 3x^2z - y^2z^3 + 4x^3y + 2x - 3y - 5$.
$[\nabla^2 f(x, y, z) = 6z + 24xy - 2z^3 - 6y^2z]$

3.9 Verify if the Laplacian of $(1/r)$ is zero, with $r = |\boldsymbol{r}| = \sqrt{x^2+y^2+z^2}$. [Yes]

3.10 Verify if the vector field: $\boldsymbol{F}(x, y, z) = (6xy + z^3)\boldsymbol{i} + (3x^2 - z)\boldsymbol{j} + (3xz^2 - y)\boldsymbol{k}$ is irrotational. [Yes]

3.11 Given the function: $f(x, y) = 4x^5 + \sin(4y)$, determine the directional derivative of $f$ in $P_0 = (1,0)$ along the unit vector:

$$\boldsymbol{u} = \frac{1}{\sqrt{10}}(-\boldsymbol{i}+3\boldsymbol{j}).$$ $\left[-\frac{8}{\sqrt{10}}\right]$

3.12 Given the function: $f(x, y) = 8y \cos x + \arctan(-8x)$, determine the directional derivative of $f$ in $P_0 = (0,1)$ along the unit vector:

$$\boldsymbol{u} = \frac{1}{\sqrt{5}}(-2\boldsymbol{i}+\boldsymbol{j}).$$ $\left[\frac{24}{\sqrt{5}}\right]$

3.13 Given the function: $f(x, y) = 2xy^3$, determine the directional derivative of $f$ in $P_0 = (-\sqrt{3},1)$ along the unit vector:

$$\boldsymbol{u} = \frac{1}{2}\boldsymbol{i} - \frac{\sqrt{3}}{2}\boldsymbol{j}.$$ [10]

3.14 Given the function: $f(x, y, z) = 4xz^3 - 3x^2y^2z$, determine the directional derivative of $f$ in $P_0 = (2,-1,2)$ along in the direction $2\boldsymbol{i} - 3\boldsymbol{j} + 6\boldsymbol{k}$. [376/7]

3.15 Given the function: $f(x, y, z) = 2z - x^3y$ and the field $\boldsymbol{F}(x, y, z) = 2x^2\boldsymbol{i} - 3yz\boldsymbol{j} + xz^2\boldsymbol{k}$ determine $\boldsymbol{F}\cdot(\nabla f)$ and $\boldsymbol{F} \times (\nabla f)$ in $P_0 = (1,-1,1)$. [5, $7\boldsymbol{i} - \boldsymbol{j} - 11\boldsymbol{k}$]

Chapter 4

# Coordinate Systems and Important Theorems

## 4.1 Orthogonal Curvilinear Coordinates

Up to this point, all calculations have been performed in rectangular Cartesian coordinates. Many calculations in science can be greatly simplified by using, instead of the familiar rectangular Cartesian coordinate system, another kind of system which takes advantage of the relations of symmetry involved in particular considered problems.

For example, if we are dealing with a sphere, it is easier to describe the position of a point on the sphere by the spherical coordinates $(r, \varphi, \theta)$. Spherical coordinates are a special case of the orthogonal curvilinear coordinate system. Let us now proceed to discuss these more general coordinate systems, in order to obtain expressions for the gradient, divergence, curl and Laplacian. Let the new coordinates $u$, $v$, $w$ be defined by specifying the Cartesian coordinates $(x, y, z)$ as functions of $(u, v, w)$:

$$x = f(u, v, w), y = g(u, v, w), z = h(u, v, w), \tag{4.1}$$

where $f$, $g$, $h$ are assumed to be continuous and differentiable. A point $P$ in space can then be defined not only by the rectangular coordinates $(x, y, z)$, but also by curvilinear coordinates $(u, v, w)$.

*Mathematics and Physics for Nanotechnology: Technical Tools and Modelling*
Paolo Di Sia

ISBN 978-981-4800-02-0 (Hardcover), 978-0-429-02775-8 (eBook)
www.panstanford.com

In the new system, $\boldsymbol{dr}$ takes the form:

$$\boldsymbol{dr}=\frac{\partial \boldsymbol{r}}{\partial u}du+\frac{\partial \boldsymbol{r}}{\partial v}dv+\frac{\partial \boldsymbol{r}}{\partial w}dw\,. \tag{4.2}$$

The vector $\partial\boldsymbol{r}/\partial u$ is tangent to the $u$ coordinate curve at $P$. If $\hat{u}$ is a unit vector at $P$ in this direction, then it is:

$$\hat{\boldsymbol{u}}=\frac{\dfrac{\partial \boldsymbol{r}}{\partial u}}{\left|\dfrac{\partial \boldsymbol{r}}{\partial u}\right|}\,. \tag{4.3}$$

We write:

$$\frac{\partial \boldsymbol{r}}{\partial u}=\left|\frac{\partial \boldsymbol{r}}{\partial u}\right|\hat{u}=h_1\,\hat{u}\,. \tag{4.4}$$

Similarly, we have:

$$\frac{\partial \boldsymbol{r}}{\partial v}=\left|\frac{\partial \boldsymbol{r}}{\partial v}\right|\hat{v}=h_2\,\hat{v}\,, \tag{4.5}$$

$$\frac{\partial \boldsymbol{r}}{\partial w}=\left|\frac{\partial \boldsymbol{r}}{\partial w}\right|\hat{w}=h_3\hat{w}\,. \tag{4.6}$$

Then $\boldsymbol{dr}$ can be written as:

$$\boldsymbol{dr}=h_1du\hat{u}+h_2dv\hat{v}+h_3dw\hat{w}\,. \tag{4.7}$$

The quantities $h_1$, $h_2$, $h_3$ are also called *scale factors*. If $\hat{u}$, $\hat{v}$, $\hat{w}$ are mutually perpendicular at any point $P$, the curvilinear coordinates are called *orthogonal*. In such a case, the element of arc length $ds$ is given by:

$$ds^2=\boldsymbol{dr}\cdot\boldsymbol{dr}=h_1{}^2du^2+h_2{}^2dv^2+h_3{}^2\,dw^2\,. \tag{4.8}$$

Along a $u$ curve, $v$ and $w$ are constants, so that $\boldsymbol{dr}=h_1du\hat{u}$. Then the differential of arc length $ds_1$ along $u$ at $P$ is $h_1du$. Similarly, the differential arc lengths along $v$ and $w$ at $P$ are $ds_2 = h_2dv$, $ds_3 = h_3dw$, respectively.

The volume of the parallelepiped is given by:

$$dV=\left|(h_1du\hat{u})\cdot(h_2dv\hat{v})\times(h_3dw\hat{w})\right|=h_1h_2h_3du\,dv\,dw\,, \tag{4.9}$$

since $\left|\hat{u}\cdot\hat{v}\times\hat{w}\right|=1$. Alternatively, $dV$ can be written as:

$$dV = \left|\frac{\partial(x,y,z)}{\partial(u,v,w)}\right| du\,dv\,dw\,, \tag{4.10}$$

where

$$J = \frac{\partial(x,y,z)}{\partial(u,v,w)} = \begin{vmatrix} \frac{\partial x}{\partial u} & \frac{\partial x}{\partial v} & \frac{\partial x}{\partial w} \\ \frac{\partial y}{\partial u} & \frac{\partial y}{\partial v} & \frac{\partial y}{\partial w} \\ \frac{\partial z}{\partial u} & \frac{\partial z}{\partial v} & \frac{\partial z}{\partial w} \end{vmatrix} \tag{4.11}$$

is called the *Jacobian* of the transformation. We assume that the Jacobian $J$ is different from zero, so that the transformation (Eq. 4.1) is one to one in the neighbourhood of a point.

## 4.2 Special Orthogonal Coordinate Systems

There are at least nine special orthogonal coordinate systems; the most common and useful ones are the cylindrical and spherical coordinates. We introduce only these two coordinates in this chapter.

### 4.2.1 Cylindrical Coordinates

The cylindrical coordinates ($R$, $\vartheta$, $z$) are useful in situations in which there is a *cylindrical* or *axial symmetry* (Fig. 4.1). From the figure, we can see that:

$$x = R\cos\vartheta, \quad y = R\sin\vartheta, \quad z = z, \tag{4.12}$$

where $R \geq 0;\ 0 \leq \vartheta < 2\pi;\ -\infty < z < \infty.$ (4.13)

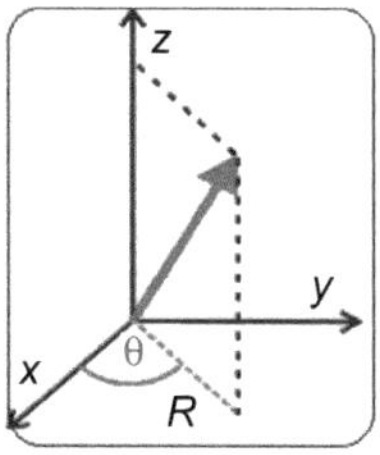


**Figure 4.1** Cylindrical coordinates.

The calculation of the Jacobian of the transformation gives:

$$\left|\frac{\partial(x,y,z)}{\partial(u,v,w)}\right| = R\,. \tag{4.14}$$

### 4.2.2 Spherical Coordinates

The spherical coordinates ($R$, $\vartheta$, $\varphi$) are useful in situations in which there is a spherical symmetry (Fig. 4.2).

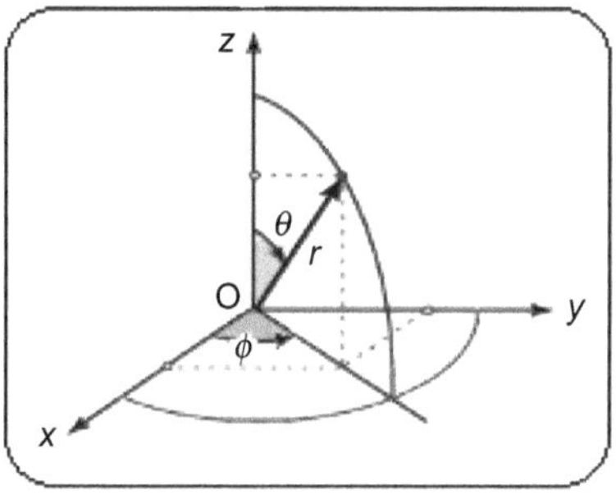


**Figure 4.2** Spherical coordinates.

From the figure, we can see that:

$$x = R\sin\vartheta\cos\varphi;\quad y = R\sin\vartheta\sin\varphi;\quad z = R\cos\vartheta. \tag{4.15}$$

where $R \geq 0;\ 0 \leq \vartheta \leq \pi;\ 0 \leq \varphi < 2\pi.$ (4.16)

The calculation of the Jacobian of the transformation gives:

$$\left|\frac{\partial(x,y,z)}{\partial(u,v,w)}\right| = R^2\sin\vartheta\,. \tag{4.17}$$

## 4.3 Vector Integration and Integral Theorems

The integration of a vector, which is a function of a single scalar $u$, can proceed as an ordinary scalar integration. For a given vector:

$$\boldsymbol{F}(u) = F_1(u)\boldsymbol{i} + F_2(u)\boldsymbol{j} + F_3(u)\boldsymbol{k}\,, \tag{4.18}$$

we have:

$$\int \boldsymbol{F}(u)du = \boldsymbol{i}\int F_1(u)du + \boldsymbol{j}\int F_2(u)du + \boldsymbol{k}\int F_3(u)du + \boldsymbol{a}\,, \tag{4.19}$$

where $\boldsymbol{a}$ is the constant of integration (a constant vector).

Now, let us consider the integral of the scalar product of a vector $\boldsymbol{a} = (x_1, x_2, x_3)$ and $d\boldsymbol{r}$ between the two points $P_1 = (x_1, x_2, x_3)$ and $P_2 = (x_4, x_5, x_6)$:

$$\int_{P_1}^{P_2} \boldsymbol{a}\cdot d\boldsymbol{r}\,. \tag{4.20}$$

It is a path-dependent integral and is called a *line integral* (or *path integral*).

If the scalar product $\boldsymbol{a} \cdot d\boldsymbol{r}$ is equal to an exact differential, i.e., $\boldsymbol{a} \cdot d\boldsymbol{r} = d\varphi$, the integration depends only upon the two points $P_1$, $P_2$ and is therefore, path-independent:

$$\int_{P_1}^{P_2} \boldsymbol{a}\cdot d\boldsymbol{r} = \int_{P_1}^{P_2} d\varphi = \varphi_2 - \varphi_1\,. \tag{4.21}$$

A vector field $\boldsymbol{a}$ which has the above property (is path-independent) is called *conservative*. Therefore, the line integral above is zero along any close path and the curl of a conservative vector field is zero:

$$d\varphi = \nabla\varphi\cdot d\boldsymbol{r} \rightarrow \nabla\times(\nabla\varphi)=0. \tag{4.22}$$

A typical example of a conservative vector field in mechanics is a conservative force.

The surface integral of a vector function $\boldsymbol{a} = (x_1, x_2, x_3)$ over the surface $S$ is an important quantity; it is defined as:

$$\int_S \boldsymbol{a}\cdot d\boldsymbol{s}\,, \tag{4.23}$$

where the surface integral stands for a double integral over a certain surface $S$ and $d\boldsymbol{s}$ is an element of area of the surface (Fig. 4.3), a vector quantity. We attribute to $d\boldsymbol{s}$ (a magnitude $ds$) and also a direction corresponding to the normal $\boldsymbol{n}$ to the surface at the point in question, thus:

$$d\boldsymbol{s} = \boldsymbol{n}ds. \tag{4.24}$$

The normal $\boldsymbol{n}$ to a surface may be taken to lie in either of two possible directions. However, if $ds$ is a part of a closed surface, the sign of $\boldsymbol{n}$ relative to $ds$ is so chosen that it points outward away from the interior.

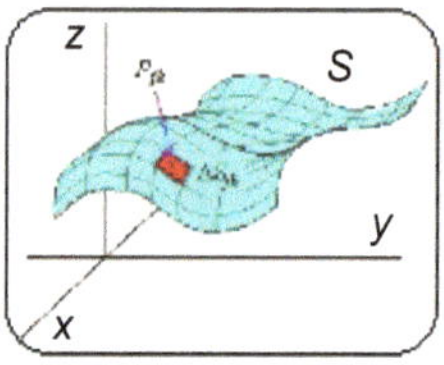


**Figure 4.3** Surface integral over a surface $S$.

If a surface integral is to be evaluated over a closed surface $S$, the integral is written as:

$$\oint_S \boldsymbol{a} \cdot \boldsymbol{ds} . \tag{4.25}$$

Note that this is different from a closed path line integral. When the path of integration is closed, the line integral is written as:

$$\oint_\gamma \boldsymbol{a} \cdot \boldsymbol{ds} , \tag{4.26}$$

where $\gamma$ specifies the closed path and $\boldsymbol{ds}$ is an element of length along the given path. By convention, $\boldsymbol{ds}$ is taken positive along the direction in which the path is traversed. Here we are only considering simple closed curves. A simple closed curve does not intersect itself anywhere.

## 4.4 Gauss Theorem

Gauss theorem, also called *theorem of divergence*, relates the surface integral of a given vector function with the volume integral of the divergence of that vector.

Let $D$ be a regular limited domain of $\mathfrak{R}^3$, whose boundary $S$ is a closed, oriented and regular surface; let $\boldsymbol{n} = \boldsymbol{n}(x,y,z)$ be the orthogonal unit vector to $S$ in the generic point $P$, outwards directed (Fig. 4.4).

Consider: $\boldsymbol{F}(x,y,z) = F_1\,(x,y,z)\,\boldsymbol{i} + F_2\,(x,y,z)\,\boldsymbol{j} + F_3\,(x,y,z)\,\boldsymbol{k}$ as a vector field $\in C^1\,(V) \subseteq \mathfrak{R}^3$, with $D \subset V$.

It results:

$$\iiint_D div\, \boldsymbol{F}(x,y,z) dxdydz = \iint_S \boldsymbol{F}(x,y,z) \cdot \boldsymbol{n}(x,y,z) ds . \tag{4.27}$$

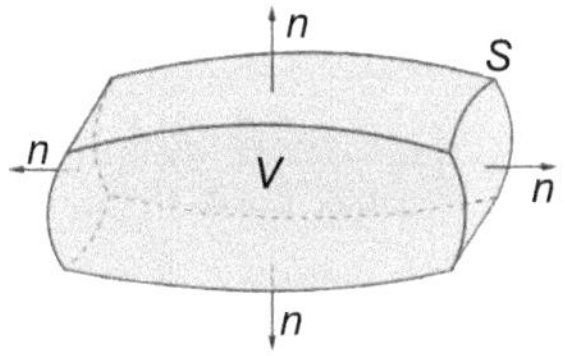


**Figure 4.4** Gauss theorem.

If we are in two dimensions, we say: let $D$ be a regular limited subset of $\Re^2$, whose boundary $C$ is a closed, oriented and regular line; let $\boldsymbol{n} = \boldsymbol{n}(x,y)$ be the orthogonal unit vector to $D$ in the generic point $P$, outwards directed.

Let us consider: $\boldsymbol{F}(x,y) = F_1\,(x,y)\,\boldsymbol{i} + F_2\,(x,y)\,\boldsymbol{j}$ as a vector field $\in C^1\,(A) \subseteq \Re^2$, with $D \subset A$. It results:

$$\iint_D div\,\boldsymbol{F}(x,y)dxdy = \oint_C \boldsymbol{F}(x,y)\cdot\boldsymbol{n}(x,y)dl\,. \tag{4.28}$$

## 4.5 Stokes Theorem

Stokes theorem relates the line integral of a vector function with the surface integral of the curl of that vector.

Let $S$ be a regular and oriented surface of $\Re^3$, whose boundary is given by a close and regular curve $C$ (or many close regular curves), with orientation inherited by the orientation of $S$ (Fig. 4.5). Let $\boldsymbol{n} = \boldsymbol{n}(x,y,z)$ be the orthogonal unit vector to $S$ in the generic point $P$ and $\boldsymbol{F}(x,y,z)$ as a vector field $\in C^1\,(V) \subseteq \Re^3$, with $S \subset V$. It results:

$$\iint_S rot\,\boldsymbol{F}(x,y,z)\cdot\boldsymbol{n}(x,y,z)ds = \oint_C \boldsymbol{F}\cdot d\boldsymbol{r}\,. \tag{4.29}$$

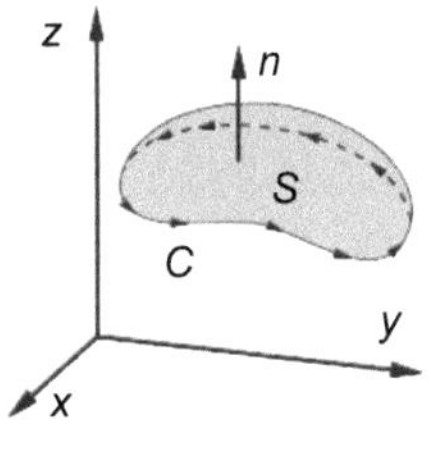


**Figure 4.5** Stokes theorem.

## 4.6 Green Theorem

*Premise*: Let $D$ be a limited subset of $\mathfrak{R}^2$, simple with respect to both Cartesian axes and with the boundary $C$ made up by the union of a finite number of regular curves. We say that $C$ is 'positively oriented' if it is fixed the counterclockwise direction of travel. With this choice, the points of $D$ remain always to the left of the boundary.

Let $C : r\ (t)\ (\ a \leq t \leq b\ )$ be a regular, simple and close curve of the plane. We indicate with $D$ the inner region to $C$ plus the support of $C$. Suppose that the travel direction of $C$ is positive, i.e., counterclockwise.

Let $\boldsymbol{F}(x,y) = F_1(x,y)\ \boldsymbol{i} + F_2(x,y)\ \boldsymbol{j}$ be a vector field $\in C^1$, defined in an open set $V$ of $\mathfrak{R}^2$, with $D \subset V$. Then it holds:

$$\oint_C \boldsymbol{F} \cdot d\boldsymbol{r} = \iint_D \left[ \frac{\partial F_2}{\partial x}(x,y) - \frac{\partial F_1}{\partial y}(x,y) \right] dx\, dy\,. \tag{4.30}$$

The Stokes theorem is the $\mathfrak{R}^3$ version of the Green theorem, when we consider the surface $S$ in the $xy$ plane.

## 4.7 Helmholtz Theorem

The divergence and curl of a vector field play very important roles in science. We learned in previous sections that a divergence-free field is solenoidal and a curl-free field is irrotational. We may classify vector fields in accordance with their being solenoidal and/or irrotational. A vector field $\boldsymbol{F}$ is:

(a) *Solenoidal and irrotational*: if $\nabla \cdot \boldsymbol{F} = 0$ and $\nabla \times \boldsymbol{F} = \boldsymbol{0}$. A static electric field in a charge-free region is a good example.

(b) *Solenoidal*: if $\nabla \cdot \boldsymbol{F} = 0$, but $\nabla \times \boldsymbol{F} \neq \boldsymbol{0}$. A steady magnetic field in a current-carrying conductor meets these conditions.

(c) *Irrotational*: if $\nabla \times \boldsymbol{F} = \boldsymbol{0}$, but $\nabla \cdot \boldsymbol{F} \neq 0$. A static electric field in a charged region is an irrotational field.

The most general vector field, such as an electric field in a charged medium with a time-varying magnetic field, is neither solenoidal nor irrotational, but can be considered as the sum of a solenoidal field and an irrotational field. This is made clear by Helmholtz theorem:

*A vector field is uniquely determined by its divergence and curl in a region of space and its normal component over the boundary of the region. In particular, if both divergence and curl are specified everywhere and if they both disappear at infinity sufficiently rapidly, then the vector field can be written as a unique sum of an irrotational part and a solenoidal part.*

We may write:

$$\boldsymbol{F}(\boldsymbol{r}) = -\nabla\,\varphi(\boldsymbol{r}) + \nabla\times\boldsymbol{A}(\boldsymbol{r}), \tag{4.31}$$

where '$-\nabla\,\varphi(\boldsymbol{r})$' is the irrotational part and '$\nabla\times\boldsymbol{A}(\boldsymbol{r})$' is the solenoidal part. $\varphi(\boldsymbol{r})$ is called *scalar potential* of $\boldsymbol{F}(\boldsymbol{r})$ and $\boldsymbol{A}(\boldsymbol{r})$ is *vector potential* of $\boldsymbol{F}(\boldsymbol{r})$. If both $\boldsymbol{A}$ and $\varphi$ can be determined, the theorem is verified.

In relation to the determination of $\boldsymbol{A}$ and $\varphi$, if the vector field $\boldsymbol{F}(\boldsymbol{r})$ is such that:

$$\nabla\cdot\boldsymbol{F}(\boldsymbol{r}) = \rho \text{ and } \nabla\times\boldsymbol{F}(\boldsymbol{r}) = \mathbf{v}, \tag{4.32}$$

then we have:

$$\nabla\cdot\boldsymbol{F}(\boldsymbol{r}) = \rho = -\nabla\cdot[\nabla\,\varphi(\boldsymbol{r})] + \nabla\cdot[\nabla\times\boldsymbol{A}(\boldsymbol{r})], \tag{4.33}$$

or

$$\nabla^2\,\varphi(\boldsymbol{r}) = -\rho \tag{4.34}$$

which is known as *Poisson equation*. Next, we have:

$$\nabla\times\boldsymbol{F}(\boldsymbol{r}) = \mathbf{v} = -\nabla\times[-\nabla\,\varphi(\boldsymbol{r}) + \nabla\times\boldsymbol{A}(\boldsymbol{r})], \tag{4.35}$$

or

$$\nabla^2\boldsymbol{A}(\boldsymbol{r}) = \boldsymbol{v} \tag{4.36}$$

or, in component, we have:

$$\nabla^2 A_i = v_i\,,\ i = 1, 2, 3 \tag{4.37}$$

where these are also Poisson equations. Thus, both $\boldsymbol{A}$ and $\varphi$ can be determined by solving Poisson equations.

## 4.8 Useful Integral Relations

These relations are closely related to the general integral theorems that we have previously considered.

(a) Integral of the gradient of a scalar field between two points $P_1$ and $P_2$:

$$\int_{P_1}^{P_2} (\nabla \varphi) \cdot d\boldsymbol{l} = \varphi(b) - \varphi(a) . \tag{4.38}$$

(b) Relation between a surface integral of a scalar field and the volume integral of the Laplacian of this field:

$$\oiint_S \frac{\partial \varphi}{\partial n} ds = \iiint_V \nabla^2 \varphi dV . \tag{4.39}$$

(c) Relation between the volume integral of the gradient of a scalar field and the surface integral of this field, with $S$ the surface of the volume $V$:

$$\iiint_V \nabla \phi dV = \oiint_S \phi \hat{n} ds . \tag{4.40}$$

(d) Relation between the volume integral of the curl of a vector field and the surface integral of this field, with $S$ the surface of the volume $V$:

$$\iiint_V \nabla \times \boldsymbol{B} dV \iint_S \hat{n} \times \boldsymbol{B} ds . \tag{4.41}$$

## Exercises

4.1 Using cylindrical coordinates, calculate:

$$\iiint_D (x^2 + y^2) dx\, dy\, dz$$

with $D = \{(x,y,z) \in \Re^3 : x^2 + y^2 \leq 4 ; x \geq 0 ; 0 \leq z \leq 3\}$. [$12\pi$]

4.2 Using spherical coordinates, calculate:

$$\iiint_D z\, dx\, dy\, dz$$

with $D = \{(x,y,z) \in \Re^2 : x^2 + y^2 + z^2 \leq 9 ; z \geq 0\}$. [$(81/4)\pi$]

4.3 Calculate the circulation of the vector field:

$\boldsymbol{F}(x, y, z) = (y + z)\boldsymbol{i} + (x + z)\boldsymbol{j} + (x - y)\boldsymbol{k}$

along the curve $C$ given by the intersection between the spherical surface $x^2 + y^2 + z^2 = 1$ and the plane $z = y$. [0]

4.4 Let us consider the vector field:

$\boldsymbol{F} = xy^2\boldsymbol{i} + x^2y\boldsymbol{j} + (x^2 + y^2)z^2\boldsymbol{k}$

with $S$ the closed surface, which delimits the solid $E$ defined by: $4 - x^2 - y^2 \geq 0$ ; $0 \leq z \leq 1$.

Determine:

$$\oiint_S \mathbf{F} \cdot d\mathbf{s}. \qquad [16\,\pi]$$

4.5 Let us consider Gauss law:

$$\int_{S\,(\text{closed})} \mathbf{E} \cdot \mathbf{n}\,da = \frac{q_{in}}{\varepsilon_0}.$$

Describe the distribution of charges by means of a charge density of the form:

$$q_{in} = \iiint_V \rho\,dV. \qquad \text{[Maxwell first equation]}$$

4.6 Let us consider in $\Re^3$ the pyramid $V$, with vertices the points (0, 0, 2), (1, 0, 0), (0, 1, 0), (−1, 0, 0), (0, −1, 0), with $S$ the total surface of $V$ and $\mathbf{n}(x,y,z)$ the orthogonal unit vector, external to $S$ in $(x,y,z)$.

Considering the vector field:

$\mathbf{F}(x, y, z) = (2x^2 - z^3)\mathbf{i} + (y^4 + 2\cos x)\mathbf{j} + (2e^y - 12z)\mathbf{k}$, calculate:

$$I = \iint_S \mathbf{F} \cdot \mathbf{n}\,ds. \qquad [-16]$$

4.7 Let $D = \{(x,y) \in \Re^2 : x^2 + y^2 \leq 4; y \geq 0\}$ and $C$ the curve-boundary of $D$, travelled once counterclockwise. Let us consider:

$$F(x,y) = \left(\frac{1}{7}x^3 y\right)\mathbf{i} + (7xy)\mathbf{j},\ \forall\ (x,y) \in \Re^2.$$

Calculate:

$$I = \oint_C \mathbf{F} \cdot d\mathbf{r}. \qquad [37{,}33]$$

4.8 Let is $D = \{(x,y) \in \Re^2 : -2 \leq x \leq 0; -5 \leq y \leq 5\}$ and $C$ the curve-boundary of $D$, travelled once counterclockwise. Let us consider:

$\mathbf{F}(x, y) = (x^2 - 2xy)\mathbf{i} + (5x^2 y)\mathbf{j},\ \forall\ (x,y) \in \Re^2$.

Calculate:

$$I = \oint_C \boldsymbol{F} \cdot d\boldsymbol{r}. \qquad [-40]$$

4.9 Let us consider in $\Re^2$ the trapezium $D$, having as vertices (in the order) the points (1, 0), (2, 1), (−2, 1), (−1, 0), with $C$ the curve-boundary of $D$ and $\boldsymbol{n}(x,y)$ the orthogonal unit vector, external to $C$ in $(x,y)$ and let:

$\boldsymbol{F}(x, y) = (9xy)\boldsymbol{i} + (x^2y^2)\boldsymbol{j},\ \forall\ (x,y) \in \Re^2$.

Calculate:

$$I = \oint_C \boldsymbol{F} \cdot \boldsymbol{n} ds. \qquad [15]$$

4.10 Let us consider in $\Re^2$ the polygon $T$, having as vertices (in the order) the points (0, 1), (1, 1), (3, 0), (1, −1), (0, −1), with $C$ the curve-boundary of $T$, travelled once counterclockwise and let:

$\boldsymbol{F}(x, y) = 3y\boldsymbol{i} + x^3y^3\boldsymbol{j},\ \forall\ (x,y) \in \Re^2$.

Calculate:

$$I = \oint_C \boldsymbol{F} \cdot d\boldsymbol{r}. \qquad [-12]$$

# Chapter 5

# Ordinary Differential Equations

## 5.1 Introduction

In science, there are a variety of reasons for studying differential equations; almost all the elementary and numerous of the advanced parts of theoretical physics (for example) are mathematically posed in terms of differential equations.

A *differential equation* is an equation that contains derivatives of an unknown function, which expresses the seek relationship. If there is only one independent variable and, as a consequence, total derivatives like $dx/dt$, the equation is called an *ordinary differential equation* (ODE). A *partial differential equation* (PDE) contains several independent variables and hence the derivative(s) in the equation are partial derivatives.

The *order* of a differential equation is the order of the highest derivative appearing in the equation.

The *degree* is the power of the derivative of the highest order after the equation has been rationalised, i.e., after fractional powers of all derivatives have been removed.

***Examples***:

(a) The equation:

*Mathematics and Physics for Nanotechnology: Technical Tools and Modelling*
Paolo Di Sia

ISBN 978-981-4800-02-0 (Hardcover), 978-0-429-02775-8 (eBook)
www.panstanford.com

$$3\frac{d^2y}{dx^2} - \frac{1}{2}\frac{dy}{dx} + 7y = 0 \tag{5.1}$$

is of second-order and first degree.

(b) The equation:

$$3\frac{d^3y}{dx^3} = \sqrt{\left(\frac{dy}{dx}\right)^3 - 5} \tag{5.2}$$

is of third order and second degree, since it contains the term $(d^3y/dx^3)^2$ after it is rationalised. We can write also the notation $y'$ for $dy/dx$, $y''$ for $d^2y/dx^2$, … , $y^{(n)}$ for $d^ny/dx^n$.

A differential equation is said to be *linear* if each term in it is such that the dependent variable or its derivatives occur only once and only to the first power.

***Examples:***

(a) The equation: $2y''' - 7y'y = 0$ (5.3)

is not linear;

(b) The equation: $x^2y''' - \sin x \cos x\, y' + 3y = 2e^x$ (5.4)

is linear.

If in a linear differential equation, there are no terms independent of $y$ (dependent variable), the equation is also said to be *homogeneous*; this would have been true for Eq. 5.4 if the $e^x$ term on the right-hand side had been replaced by zero.

A very important property of linear homogeneous equations is that, if we know two solutions $y_1$ and $y_2$, we can construct others as linear combinations of them; this is known as the *principle of superposition*. Sometimes differential equations look unfamiliar. A trivial change of variables can reduce a seemingly impossible equation into one whose type is readily recognisable.

Many differential equations are very difficult to solve. There are only a relatively small number of types of differential equation that can be solved in closed form. We start with equations of *first-order*. A first-order differential equation can always be solved, although the solution may not always be expressible in terms of familiar functions.

A *solution* (or *integral*) of a differential equation is the relation between the variables, not involving differential coefficients, which satisfies the differential equation. The solution of a differential equation of order $n$ in general involves $n$ arbitrary constants.

***First-order differential equations***: A first-order differential equation has the general form as follows:

$$\frac{dy}{dx} = \frac{f(x,y)}{g(x,y)}. \tag{5.5}$$

## 5.2 Separable Variables

If $f(x,y)$ and $g(x,y)$ are reducible to $P(x)$ and $G(y)$, respectively, then we have:

$$P(x)dx = G(y)dy. \tag{5.6}$$

Its solution is found at once by integrating.

***Examples:***

(a) We consider the equation: $y' = x$. It can be written as:

$dy/dx = x$

or $dy = x\,dx$.

Integrating both members, we have:

$y + c_1 = 1/2\,x^2 + c_2,$

or $y = 1/2\,x^2 + c$.

(b) We consider the equation: $y' = 2\,x$. It can be written as:

$dy/dx = 2\,x$

or $dy = 2\,x\,dx$.

Integrating both members, we have:

$y = x^2 + c$.

(c) We consider the equation: $y' + y^2\,e^x = 0$. It can be written as:

$dy/y^2 = -\,e^x\,dx$.

Integrating both members, we have:

$1/y = e^x + c$.

*Note*: The various constants appearing in both members, also multiplied or divided by numerical factors, can be written as a single constant $c$.

## 5.3 First-Order Linear Equation

A first-order linear equation is writable in the form:

$$y'(x) + P(x)y(x) = Q(x). \tag{5.7}$$

The associated homogeneous equation is:

$$y'(x) + P(x)y(x) = 0. \tag{5.8}$$

By separation of the variables, Eq. 5.8 brings to the solution:

$$y(x) = ce^{-\int P(x)dx}. \tag{5.9}$$

The solution of Eq. 5.7 is:

$$y(x) = e^{-\int P(x)dx}\left[c + \int Q(x)e^{\int P(x)dx}dx\right]. \tag{5.10}$$

So, if we recognise a first-order linear equation, we can apply directly the expression of Eq. 5.10.

*Examples:*

(a) Let us consider the equation: $y' = x\,(2 + y)$.

It can be written as: $y' - xy = 2x$. We recognise $P(x) = -x$ and $Q(x) = 2x$. The application of Eq. 5.10 gives:

$$y(x) = ce^{\frac{1}{2}x^2} - 2.$$

It is a family of infinite curves in $\Re^2$, with $c \in \Re$. For example, for $c = 1$, we have the curve in Fig. 5.1.

(b) Let us consider the amount of substance, which is growing or decaying. Calling $N(t)$ this amount, let us consider the time rate of change proportional to the amount of substance. Study the followed differential equation.

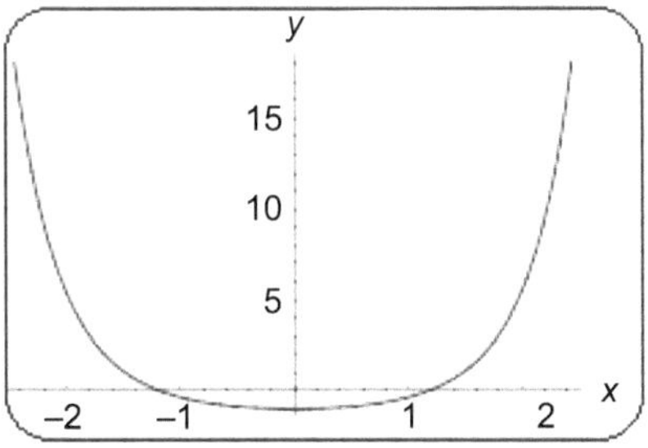


**Figure 5.1** Plot of previous solution with $c = 1$.

We can write $\dfrac{dN(t)}{dt} = kN$.

This differential equation is both linear and separable; the solution is:

$$\frac{dN(t)}{N(t)} = k\,dt.$$

Integrating both members, we have:

$N(t) = N_0 e^{k(t-t_0)}$,

where $N_0$ is the substance at the initial time $t_0$. This is the differential equation for the *radioactive decay* and for *population growth.* If the constant $k$ is negative, we have a decay equation and if $k$ is positive, we have a growth equation.

## 5.4 Bernoulli Equations

This equation is a non-linear first-order equation, which can be reduced to a linear one by a substitution. Its form is:

$$y'(t) = a(t)y(t) + b(t)y^{\alpha}(t), \tag{5.11}$$

with $\alpha \in \Re - \{0, 1\}$.

Indeed: if $\alpha = 0$, Eq. 5.11 becomes a first-order linear differential equation; if $\alpha = 1$, Eq. 5.11 becomes a homogeneous first-order linear differential equation.

Dividing both members for $y^{\alpha}$, writing $y^{1-\alpha}(t) = z(t)$, deriving this last for getting $z'(t)$, we find a linear first-order equation in the variable $z$. Then we return to $y$ variable.

***Example* 5.1:** Solve the equation: $y'(t) = y(t) - t\,y^2(t)$, with the initial condition: $y(0) = 1$.

***Solution*:** It is a Bernoulli equation with $\alpha = 2$. Following the above indicated procedure, we divide both members for $y^2$, put: $y^{-1}(t) = z(t)$ and derive this last, obtaining: $z'(t) = -y'(t)/y^2(t)$. Substituting the found results, we have now the equation: $z'(t) + z(t) = t$, with initial condition: $z(0) = 1$, which is an equation of the form given in Eq. 5.7. The use of Eq. 5.10 brings to: $z(t) = 2\,e^{-t} + t - 1$, i.e.,

$$y(t) = 1/z(t) = 1/(2\,e^{-t} + t - 1) \text{ (Fig. 5.2).}$$

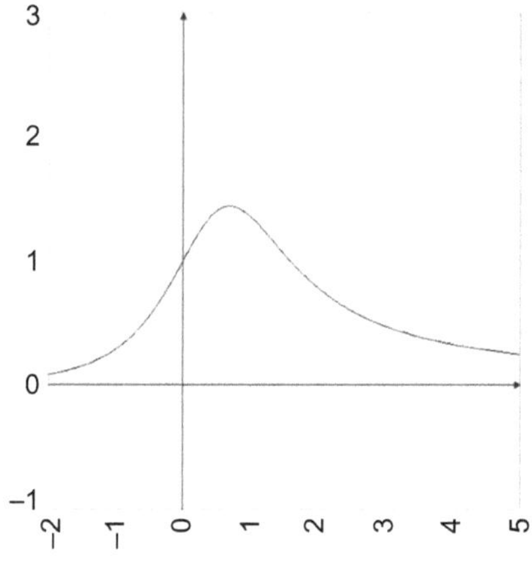


**Figure 5.2** Plot for solution of Example 5.1.

## 5.5 Second-Order Linear Equations with Constant Coefficients

The general form of a *second-order* linear differential equation is:

$$y'' + p(x)y' + q(x)y = r(x), \tag{5.12}$$

i.e., the variable $y$ and its derivatives appear with maximum exponent equal to one. If $p(x)$ and $q(x)$ are constants, we have linear equations with constant coefficients.

### 5.5.1 Homogeneous Linear Equations with Constant Coefficients

We start considering *homogeneous* linear equations with constant coefficients, of the form:

$$y''(t) + a_1 y'(t) + a_0 y(t) = 0, \tag{5.13}$$

with $a_1$, $a_0 \in \Re$. For resolving it, we write the *characteristic polynomial* associated to the equation: $P(\lambda) = \lambda^2 + a_1 \lambda + a_0$, therefore, we calculate the roots of $P(\lambda)$.

There are three possible cases as follows:

(a) *Two real different solutions* $\lambda_1$, $\lambda_2$: the solution of Eq. 5.13 is:

$$y(t) = c_1 e^{\lambda_1 t} + c_2 e^{\lambda_2 t}. \tag{5.14}$$

(b) *Two real coincident solutions* $\lambda_0$: the solution of Eq. 5.13 is:

$$y(t) = c_1 e^{\lambda_0 t} + t c_2 e^{\lambda_0 t}. \tag{5.15}$$

(c) *Two complex conjugated solutions* $\alpha + i\beta$, $\alpha - i\beta$: the solution of Eq. 5.13 is:

$$y(t) = c_1 e^{\alpha t}\cos(\beta t) + c_2 e^{\alpha t}\sin(\beta t). \tag{5.16}$$

***Example* 5.2**: Solve the equation: $y''(t) + 5\,y'(t) + 6\,y(t) = 0$, with the initial conditions: $y\,(0) = 1, y'\,(0) = 0$.

***Solution***: The characteristic polynomial is: $P(\lambda) = \lambda^2 + 5\,\lambda + 6$, with roots $\lambda_1 = -3$, $\lambda_2 = -2$.

The solution is:

$$y(t) = c_1 e^{-3t} + c_2 e^{-2t}.$$

With the initial conditions, we find the constants $c_i$. It is: $c_1 = -2$, $c_2 = 3$. The solution of equation is therefore (Fig. 5.3):

$$y(t) = -2e^{-3t} + 3e^{-2t}.$$

***Example* 5.3**: Solve the equation: $y''(t) + 6\,y'(t) + 9\,y(t) = 0$.

***Solution***: The associated characteristic polynomial in this case is: $P(\lambda) = \lambda^2 + 6\,\lambda + 9$, which has two equal real roots $\lambda_{1,2} = -3$.

The solution is:

$$y(t) = c_1 e^{-3t} + t c_2 e^{-3t}, \text{ with } c_1, c_2 \in \Re.$$

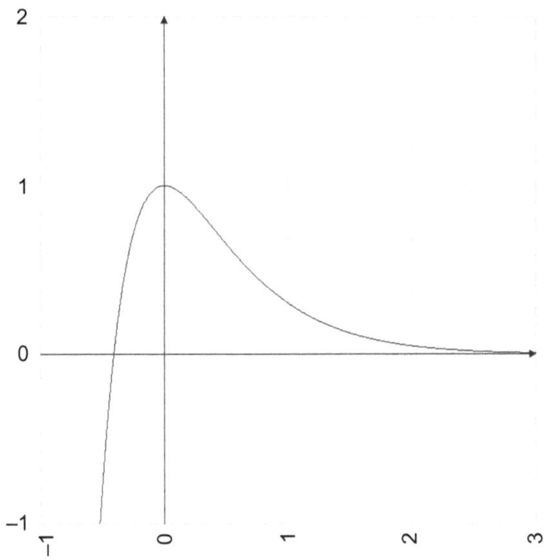


**Figure 5.3** Plot for solution of Example 5.2

***Example* 5.4:** Solve the equation: $y''(t) + 2\,y'(t) + 2\,y(t) = 0$.

***Solution*:** The associated characteristic polynomial is: $P(\lambda) = \lambda^2 + 2\lambda + 2$, which has two complex conjugate roots $\lambda_{1,2} = -1 \pm i$ or $\alpha = -1$, $\beta = 1$.

The solution in this case is:

$$y(t) = c_1 e^{-t}\cos(t) + c_2 e^{-t}\sin(t), \text{ with } c_1, c_2 \in \Re.$$

### 5.5.2 Non-homogeneous Linear Equations with Constant Coefficients

We consider now *non-homogeneous* linear equations with constant coefficients, of the form:

$$y''(t) + a_1 y'(t) + a_0\, y(t) = g(t), \tag{5.17}$$

with $a_1, a_0 \in \Re$ and $g(t)$ a continuous function in its domain.

The below indicated procedure of resolution, known as *method of similarity*, holds for non-homogeneous linear differential equations, with constant coefficients, having a particular type of known term $g(t)$. It is as follows:

(a) We find *the solution* of the associated homogeneous equation $y_H(t)$;

(b) We find a *particular solution* of the non-homogeneous equation $\bar{y}(t)$;

(c) The general solution of Eq. 5.17 is given by:

$$y(t) = y_H(t) + \bar{y}(t). \tag{5.18}$$

The known term $g(t)$ can have the following forms:

(i) $g(t) = Q(t)$, where $Q(t)$ is a polynomial of variable $t$;

(ii) $g(t) = e^{\lambda t} Q(t)$.

In these two cases, we have three possibilities:

- It is $\lambda = \lambda_1$ or $\lambda = \lambda_2$, i.e., $\lambda$ coincides with one of the two roots of the characteristic polynomial, then the particular solution $\bar{y}(t)$ of the non-homogeneous equation is of the type:

  $\bar{y}(t) = te^{\lambda t}\bar{Q}(t)$,

  where $\bar{Q}(t)$ is a polynomial of the same degree of $Q(t)$.

- It is $\lambda = \lambda_0$, i.e., $\lambda$ coincides with the multiple root of the characteristic polynomial, then the particular solution $\bar{y}(t)$ of the non-homogeneous equation is of the type:

  $\bar{y}(t) = t^2 e^{\lambda t}\bar{Q}(t)$,

  where $\bar{Q}(t)$ is a polynomial of the same degree of $Q(t)$.

- If $\lambda$ does not coincide with the roots of the characteristic polynomial, then the particular solution $\bar{y}(t)$ of the non-homogeneous equation is of the type:

  $\bar{y}(t) = e^{\lambda t}\bar{Q}(t)$,

  where $\bar{Q}(t)$ is a polynomial of the same degree of $Q(t)$.

(iii) $g(t) = \sin(\beta t)Q(t)$;

(iv) $g(t) = \cos(\beta t)Q(t)$.

In these two cases, we have two possibilities:

- If $i\beta$ is a root of the characteristic polynomial, then the particular solution $\bar{y}(t)$ of the non-homogeneous equation is of the type:

  $\bar{y}(t) = t(\cos(\beta t)\bar{Q}(t) + \sin(\beta t)R(t))$, where $\bar{Q}(t)$ and $R(t)$ are polynomials of the same degree of $Q(t)$.

- If $i\beta$ is not a root of the characteristic polynomial, then the particular solution $\bar{y}(t)$ of the non-homogeneous equation is of the type:

  $\bar{y}(t) = \cos(\beta t)\bar{Q}(t) + \sin(\beta t)R(t)$,

  where $\bar{Q}(t)$ and $R(t)$ are polynomials of the same degree of $Q(t)$.

(v) $g(t) = e^{\alpha t}\sin(\beta t)Q(t)$;

(vi) $g(t) = e^{\alpha t}\cos(\beta t)Q(t)$.

In these two cases, we have two possibilities:

- If $\alpha = \alpha_1$ and $\beta = \beta_1$, with $\alpha_1$ and $\beta_1$ terms of the complex roots of the characteristic polynomial, then the particular solution $\bar{y}(t)$ of the non-homogeneous equation is of the type:

  $\bar{y}(t) = te^{\alpha t}(\cos(\beta t)\bar{Q}(t) + \sin(\beta t)R(t))$,

  where $\bar{Q}(t)$ and $R(t)$ are polynomials of the same degree of $Q(t)$.

- If $\alpha$ and $\beta$ do not coincide (respectively) with $\alpha_1$ and $\beta_1$, terms of the complex roots of the characteristic polynomial, then the particular solution $\bar{y}(t)$ of the non-homogeneous equation is of the type:

  $\bar{y}(t) = e^{\alpha t}[\cos(\beta t)\bar{Q}(t) + \sin(\beta t)R(t))$,

  where $\bar{Q}(t)$ and $R(t)$ are polynomials of the same degree of $Q(t)$.

(vii) Algebraic sum of one or more of the previous terms.

***Example* 5.5:** Solve the equation: $y''(t) + 3\,y\,(t) = 6 - t$.

***Solution*:** In this case, $Q(t)$ (right-hand side) is a first-grade polynomial of variable $t$. In the left-hand side, the term of higher order is $y$ (first-order) because the second derivative has lower order. Considering the equivalence of the two terms, we write $y$ in the form:

$$y = A\,t + B$$

or $y' = A$

or $y'' = 0$.

The initial equation assumes the form:

$$0 + 3\,(A\,t + B) = 6 - t,$$

or $A = -1/3;\ B = 2,$

or $\bar{y}(t) = -1/3t + 2$.

Now, we find the general solution of the associated homogeneous equation; the associated characteristic polynomial is:

$$P(\lambda) = \lambda^2 + 3,$$

which has two complex roots: $\lambda_{1,2} = \pm i\sqrt{3}$,
or the general solution of the associated homogeneous equation is:

$$y_H(t) = c_1 \cos(\sqrt{3}t) + c_2 \sin(\sqrt{3}t).$$

The general solution of Eq. 5.18 is:

$$y(t) = c_1 \cos(\sqrt{3}t) + c_2 \sin(\sqrt{3}t) - \frac{1}{3}t + 2, \text{ with } c_1, c_2 \in \Re.$$

***Example* 5.6:** Solve the equation: $y''(t) + y'(t) = 1 - t^2$.

***Solution*:** No $y$ appears explicitly; the highest-grade term in the left-hand side is $y'$, whose grade has to be equal to two (the grade of polynomial on the right-hand side). Therefore, we look at a particular solution of third degree:

$$\bar{y}(t) = at^3 + bt^2 + ct + d.$$

It is $d = 0$, because no $y$ appears in the equation. We calculate the first and second derivatives of $\bar{y}(t)$ and insert them in the equation. Therefore, we impose the equivalence of the two polynomials, degree by degree, obtaining: $a = -1/3$; $b = 1$; $c = -1$.

$$\Rightarrow \quad \bar{y}(t) = -\frac{1}{3}t^3 + t^2 - t.$$

Solving the associated homogeneous equation, we get the general solution:

$$y_H(t) = c_1 + c_2 e^{-t}.$$

The general solution of the initial equation is therefore:

$$y_H(t) = c_1 + c_2e^{-t} - \frac{1}{3}t^3 + t^2 - t \text{, with } c_1, c_2 \in \Re.$$

***Example* 5.7:** Solve the equation: $y''(t) - y(t) = 2t\sin(t)$.

***Solution*:** The associated homogeneous equation: $y''(t) - y(t) = 0$ has characteristic polynomial: $P(\lambda) = \lambda^2 - 1$; the general solution is:

$$y_H(t) = c_1e^t + c_2e^{-t}.$$

We search now a particular solution of the form:

$$\bar{y}(t) = A\cos(\beta t) + \sin(\beta t)R(t),$$

with $R(t)$ polynomial of degree one and $A$ constant. Then it is:

$$\bar{y}(t) = A\cos(t) + (Ct + D)\sin(t).$$

Now, we derive twice and substitute in the initial equation, obtaining $A = -1$; $C = -1$; $D = 0$. Therefore, the general solution of the initial equation is given by:

$$y(t) = c_1e^t + c_2e^{-t} - \cos(t) - t\sin(t).$$

## 5.6 An Introduction to Differential Equations with Order $k > 2$

Let us consider a linear differential equation with constant coefficients and order $k > 2$. We start with the homogeneous case.

A linear homogeneous differential equation with constant coefficients and order $k > 2$ has the following form:

$$y^{(k)}(t) + a_{k-1}y^{(k-1)}(t) + a_{k-2}y^{(k-2)}(t) + \ldots + a_1y'(t) = a_0y(t) = 0 \quad (5.19)$$

with $a_i \in \Re$. For solving Eq. 5.19:

(a) we write the associated characteristic polynomial $P(\lambda)$;

(b) we find the roots of $P(\lambda)$. There are three possibilities:

(i) $P(\lambda)$ has $k$ real solutions, all with multiplicity equal to 1: $\lambda_1, \lambda_2, \ldots, \lambda_k$.

The general solution of Eq. 5.19 is:

$$y(t) = c_1e^{\lambda_1 t} + c_2e^{\lambda_2 t} + \cdots + c_ke^{\lambda_k t} \text{, with } c_1, c_2, \ldots, c_k \in \Re.$$

(ii) $P(\lambda)$ has $r$ real solutions $\lambda_1, \lambda_2, \ldots, \lambda_r$ with multiplicity equal to 1 and $2s$ complex solutions $\mu_1, \bar{\mu}_1, \mu_2, \bar{\mu}_2, \ldots, \mu_s, \bar{\mu}_s$, with:

$\mu_j = \alpha_j + i\beta_j, \bar{\mu}_j = \alpha_j - i\beta_j, \forall j \in \{1, 2, \ldots, s\}$ and $r + 2s = k$.

The general solution of Eq. 5.19 is:

$$y(t) = c_1 e^{\lambda_1 t} + c_2 e^{\lambda_2 t} + c_r e^{\lambda_r t} + \qquad (r)$$

$$+ h_1 e^{\alpha_1 t}\cos(\beta_1 t) + h'_1 e^{\alpha_1 t}\sin(\beta_1 t) + \ldots.. \qquad (s)$$

$$\ldots.. + h_s e^{\alpha_s t}\cos(\beta_s t) + h'_s e^{\alpha_s t}\sin(\beta_s t), \qquad (s)$$

with $c_1, c_2, \ldots, c_r, h_1, h'_1, \ldots, h_s, h'_s \in \Re$.

(iii) $P(\lambda)$ has $r$ real solutions $\lambda_1, \lambda_2, \ldots, \lambda_r$ with possible multiplicity and $2s$ complex solutions $\mu_1, \bar{\mu}_1, \mu_2, \bar{\mu}_2, \ldots, \mu_s, \bar{\mu}_s$, with:

$\mu_j = \alpha_j + i\beta_j, \bar{\mu}_j = \alpha_j - i\beta_j, \forall j \in \{1, 2, \ldots, s\}$, $r + 2s = k$ and:

$\lambda_1$ with multiplicity $m_1$

$\lambda_2$ with multiplicity $m_2$

.....

$\lambda_r$ with multiplicity $m_r$

$\mu_1, \bar{\mu}_1$ with multiplicity $n_1$

$\mu_2, \bar{\mu}_2$ with multiplicity $n_2$

.....

$\mu_s, \bar{\mu}_s$ with multiplicity $n_s$

with $m_1 + m_2 + \ldots + m_r + 2n_1 + 2n_2 + 2n_s = k$, it is in general:

- If $\lambda_0 \in \Re$ is solution of $P(\lambda)$ with multiplicity $m$, then:

  $e^{\lambda_0 t}, te^{\lambda_0 t}, t^2 e^{\lambda_0 t}, \ldots, t^{m-1} e^{\lambda_0 t}$

  are $m$ linearly independent solutions of the differential equation, associated to the root $\lambda_0$.
- If $\mu = \alpha + i\beta$ and $\bar{\mu} = \alpha - i\beta$ are solution of $P(\lambda)$ with multiplicity $n$, then:

  $e^{\alpha t}\cos(\beta t), te^{\alpha t}\cos(\beta t), \ldots.. t^{n-1} e^{\alpha t}\cos(\beta t)$

  and

  $e^{\alpha t}\sin(\beta t), te^{\alpha t}\sin(\beta t), \ldots.. t^{n-1} e^{\alpha t}\sin(\beta t)$

  are $2n$ linearly independent solutions of the differential equation, associated to the roots $\mu, \bar{\mu}$. Consequently, based on the multiplicity of every solution, the general solution of the differential equation is built.

***Example* 5.8**: Solve the equation: $y^{(3)}(t) - 5y''(t) + 4y'(t) = 0$.

***Solution***: The associated characteristic polynomial $P(\lambda) = \lambda^3 - 5\lambda^2 + 4\lambda$ allows the following solutions: $\lambda_1 = 0$, $\lambda_2 = 1$, $\lambda_3 = 4$, all with multiplicity equal to 1. We are in case (i). The general solution is therefore:

$$y(t) = c_1 + c_2 e^t + c_3 e^{4t},$$

with $c_1, c_2, c_3 \in \Re$.

***Example* 5.9:** Solve the equation: $y^{(4)}(t) + y''(t) = 0$.

***Solution***: The associated characteristic polynomial $P(\lambda) = \lambda^4 + \lambda^2$ allows the solutions:

- $\lambda_1 = 0$, with multiplicity equal to 2;
- $\mu_1 = i$, $\bar{\mu}_1 = -i$, with multiplicity equal to 1.

We are in case (iii). We have:

- The solutions associated to $\lambda_1 = 0$ (with multiplicity $m = 2$) are

  $e^{\lambda_1 t} = 1; te^{\lambda_1 t} = t$.
- The solution associated to $\mu_1 = i$, i.e., $\alpha = 0$, $\beta = 1$, is

  $e^{\alpha t}\cos(\beta t) = \cos(t)$.
- The solution associated to $\bar{\mu}_1 = -i$, i.e., $\alpha = 0$, $\beta = 1$, is

  $e^{\alpha t}\sin(\beta t) = \sin(t)$.

  The general solution is therefore:

  $y(t) = c_1 + c_2 t + c_3\cos(t) + c_4\sin(t)$, with $c_1, c_2, c_3, c_4 \in \Re$.

## Exercises

5.1 Solve the equation $y''(t) - y'(t) = t$.

$$\left[y(t) = c_1 + c_2 e^t - \frac{1}{2}t^2 - t\right]$$

5.2 Solve the equation $(x^2 + 1)\,y'(x) + y^2(x) = 0$.

$$\left[y(x) = \frac{1}{\text{arc } \tan x + c}\right]$$

5.3 Solve the equation $y'(t) = \frac{ty}{(t-1)^2}$, with the initial condition:

$y(2) = 1$.

$$\left[y(t) = (t-1)\exp\left(\frac{t-2}{t-1}\right)\right]$$

5.4 Determine $a$ for which $y(x) = xe^{ax}$ is a solution of:

$x\,y'' - x\,y' - y = 0$. $[a = 1]$

5.5 Solve the equation $y''(t) + 2\,y'(t) + y(t) = \sin t$, with the initial conditions: $y(0) = 1/2, y'(0) = 1$.

$$\left[y(t) = e^{-t} + 2te^{-t} - \frac{1}{2}\cos(t)\right]$$

5.6 Solve the equation $y''(t) - 2y'(t) - 3y(t) = e^{4t}$, with the initial conditions: $y(0) = 0, y'(0) = 1$.

$$\left[y(t) + \frac{1}{5}e^{4t} - \frac{1}{5}e^{-t}\right]$$

5.7 Solve the equation $y''(t) + 2\,y'(t) + 2\,y(t) = \sin(t) + \cos(t)$.

$$\left[y(t) = e^{-t}(c_1\cos(t) + c_2\sin(t)) - \frac{1}{5}\cos(t) + \frac{3}{5}\sin(t)\right]$$

5.8 Solve the equation $y''(t) + 4\,y(t) = t + \cos(2t)$.

$$\left[y(t) = c_1\cos(2t) + c_2\sin(2t) + \frac{1}{4}t + \frac{1}{16}(\cos(2t) + 4t\sin(2t))\right]$$

5.9 Solve the equation $y' = \frac{1}{\sqrt{x}}y + 1$.

$$\left[y(x) = -\sqrt{x} - \frac{1}{2} + \frac{1}{2}ce^{2\sqrt{x}}\right]$$

5.10 Solve the equation $y''' - y = 0$, with the initial conditions:

$y(0) = 1; y'(0) = y''(0) = 0$.

$$\left[y(x) = \frac{1}{3}e^{x} + \frac{2}{3}e^{-\frac{x}{2}}\cos\left(\frac{\sqrt{3}}{2}x\right)\right]$$

# Chapter 6

# Fourier Series and Integrals

## 6.1 Periodic Functions

A Fourier series is an infinite series of sines and cosines, capable of representing almost any periodic function whether continuous or not. Periodic functions that occur in scientific problems are often very complicated and it is desirable to represent them in terms of simple periodic functions. Therefore, the study of Fourier series is a matter of great practical importance for science.

If the function $f(x)$ is defined for all $x$ and there is some positive constant $A$ such that $f(x + A) = f(x)$, then we can say that $f(x)$ is *periodic* with a *period* $A$ (Fig. 6.1).

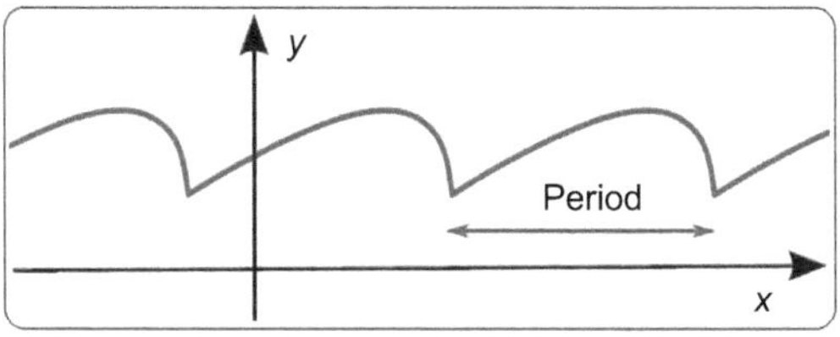


**Figure 6.1** Periodic function.

We also have, for all $x$ and any integer $n$, $f(x + nA) = f(x)$, i.e., every periodic function has arbitrarily large periods and contains


*Mathematics and Physics for Nanotechnology: Technical Tools and Modelling*
Paolo Di Sia



ISBN 978-981-4800-02-0 (Hardcover), 978-0-429-02775-8 (eBook)
www.panstanford.com

arbitrarily large numbers in its domain. We call $A$ the 'fundamental period', or simply the period.

A periodic function does not need to be defined for all values of its independent variable.

*Example*: The period of $\sin(x)$ is $2\pi$, since $\sin(x) = \sin(x + 2\pi) = \sin(x + 4\pi) = \sin(x + 6\pi) = ....$

## 6.2 Fourier Series

If the general periodic function $f(x)$ is defined in an interval $-\pi \leq x \leq \pi$, the Fourier series of $f(x)$ in $[-\pi, \pi]$ is defined to be a trigonometric series of the form:

$$f(x) = 1/2a_0 + a_1\cos(x) + a_2\cos(2x) + ..... + a_n\cos(nx) + .....$$

$$+ b_1\sin(x) + b_2\sin(2x) + ..... + b_n\sin(nx) + ..... \tag{6.1}$$

where the numbers $a_i$ and $b_i$ are called the *Fourier coefficients* of $f(x)$ in $[-\pi, \pi]$. If this expansion is possible, our power to solve physical problems is greatly increased, since the sine and cosine terms in the series can be individually handled without any difficulty. Now, let us assume that the series exists, converges and may be integrated term-by-term.

## 6.3 Euler–Fourier Formulas

Multiplying both sides of Eq. 6.1 by $\cos(mx)$, then integrating the result from $-\pi$ to $\pi$ and using important properties of sines and cosines, such as:

$$\int_{-\pi}^{\pi}\sin(mx)\sin(nx)dx = \int_{-\pi}^{\pi}\cos(mx)\cos(nx)dx = \begin{matrix} 0 & \text{if} & n \neq m \\ \pi & \text{if} & n = m \end{matrix}, \tag{6.2}$$

$$\int_{-\pi}^{\pi}\sin(mx)\cos(nx)dx = 0, \text{for all } m, n > 0, \tag{6.3}$$

we find

$$a_n = \frac{1}{\pi}\int_{-\pi}^{\pi} f(x)\cos(nx)dx\,, \tag{6.4}$$

with $n$ integer number. So, multiplying both sides of Eq. 6.1 by $\sin(mx)$ and considering the same procedure, we find:

$$b_n = \frac{1}{\pi}\int_{-\pi}^{\pi} f(x)\sin(nx)dx\,. \tag{6.5}$$

Equations 6.4 and 6.5 are known as the *Euler–Fourier formulas*. From the definition of a definite integral, it follows that, if $f(x)$ is single-valued and continuous within the interval $[-\pi, \pi]$ or merely piecewise continuous (continuous except at finite numbers of finite jumps in the interval), the integrals in Eqs. 6.4 and 6.5 exist and we may compute the Fourier coefficients of $f(x)$. If there exists a finite discontinuity in $f(x)$ at the point $x_0$, the coefficients $a_i$ and $b_i$ are determined by integrating first to $x = x_0$ and then from $x_0$ to $\pi$, as:

$$a_n = \frac{1}{\pi}\left[\int_{-\pi}^{x_0} f(x)\cos(nx)dx + \int_{x_0}^{\pi} f(x)\cos(nx)dx\right]; \tag{6.6}$$

$$b_n = \frac{1}{\pi}\left[\int_{-\pi}^{x_0} f(x)\sin(nx)dx + \int_{x_0}^{\pi} f(x)\sin(nx)dx\right]. \tag{6.7}$$

This procedure may be extended to any finite number of discontinuities.

## 6.4 Half-Range Fourier Series

Unnecessary work in determining Fourier coefficients of a function can be avoided if the function is 'odd' or 'even'. A function $f(x)$ is called *odd* if $f(-x) = -f(x)$ and *even* if $f(-x) = f(x)$. It is easy to show that in the Fourier series corresponding to an odd function $f_{\text{odd}}(x)$, only sine terms can be present in the series expansion in the interval $(-\pi, \pi)$ and it is:

$$a_n = 0; \tag{6.8}$$

$$b_n = \frac{2}{\pi}\left[\int_0^{\pi} f_{\text{odd}}(x)\sin(nx)dx\right]. \tag{6.9}$$

Similarly, in the Fourier series corresponding to an even function $f_{\text{even}}(x)$, only cosine terms (and possibly a constant) can be present. It is:

$$a_n = \frac{2}{\pi}\left[\int_0^{\pi} f_{\text{even}}(x)\cos(nx)dx\right]; \tag{6.10}$$

$$b_n = 0. \tag{6.11}$$

The Fourier coefficients $a_n$ and $b_n$ are computed in the interval $(0, \pi)$, which is half of the interval $(-\pi, \pi)$. Thus, the Fourier sine or cosine series in this case is often called a *half-range Fourier series.*

Any arbitrary function (neither even nor odd) can be expressed as a combination of $f_{\text{even}}(x)$ and $f_{\text{odd}}(x)$ as:

$$f(x) = 1/2[f(x) + f(-x)] + 1/2[f(x) - f(-x)] = f_{\text{even}}(x) + f_{\text{odd}}(x). \tag{6.12}$$

When a half-range series corresponding to a given function is desired, the function is generally defined in the interval $(0, \pi)$ and then it is specified as odd or even, so that it is clearly defined in the other half of the interval $(-\pi, 0)$.

## 6.5 Change of Interval

A Fourier expansion is not restricted to intervals as $(-\pi, \pi)$ and $(0, \pi)$. In many problems, the period of the function to be expanded may be some other interval, for example, $2L$. In this case, the problem is not a difficult one; basically all that is involved is to change the variable.

Let $z = (\pi/L)x$, then:

$$f(z) = f(\pi x/L) = F(x). \tag{6.13}$$

Thus, if $f(z)$ is expanded in the interval $-\pi < z < \pi$, the coefficients for the expansion of $F(x)$ in the interval $-L < x < L$ may be obtained by substituting Eq. 6.13 into these expressions. We have then:

$$a_n = \frac{1}{L}\int_{-L}^{L} F(x)\cos\left(\frac{n\pi}{L}\right)dx, \; n = 0, 1, 2, 3, \ldots.. \tag{6.14}$$

$$b_n = \frac{1}{L}\int_{-L}^{L} F(x)\sin\left(\frac{n\pi}{L}\right)dx \text{ , } n = 1, 2, 3, \ldots.. \tag{6.15}$$

The possibility of expanded functions in which the period is other than $2\pi$ increases the usefulness of Fourier expansion.

*Example*: Let us consider the value of $L$, it is obvious that the larger the value of $L$, the larger the basic period of the function being expanded. As $L\to\infty$, the function would not be periodic at all.

## 6.6 Parseval's Identity

Parseval's identity states that:

$$\frac{1}{2L}\int_{-L}^{L}[F(x)]^2 dx = \left(\frac{a_0}{2}\right)^2 + \frac{1}{2}\sum_{n=1}^{\infty}\left(a_n^{\ 2} + b_n^{\ 2}\right), \tag{6.16}$$

if $a_n$ and $b_n$ are coefficients of the Fourier series of $f(x)$ and if $f(x)$ satisfies the Dirichlet conditions.

Parseval's identity shows a relation between the average of the square of $f(x)$ and the coefficients in the Fourier series for $f(x)$. It holds:

(a) The average of $[f(x)]^2$ is $(1/2)\int_{-L}^{L}[f(x)]^2 dx$;

(b) The average of $(a_0/2)$ is $(a_0/2)^2$;

(c) The average of $[a_n \cos(nx)]$ is $a_n^{\ 2}/2$;

(d) The average of $[b_n \sin(nx)]$ is $b_n^{\ 2}/2$.

***Example* 6.1:** Expand $f(x) = x$, with $0 < x < 2$, in a half-range cosine series, then write Parseval's identity corresponding to this Fourier cosine series.

***Solution***: We first extend the definition of $f(x)$ to that of the even function of period 4 (Fig. 6.2).

Then $2L = 4$, so $L = 2$. Thus, $b_n = 0$ and

$$a_n = -\frac{4}{n^2\pi^2}\left(\cos(n\pi) - 1\right) \text{, if } n \neq 0.$$

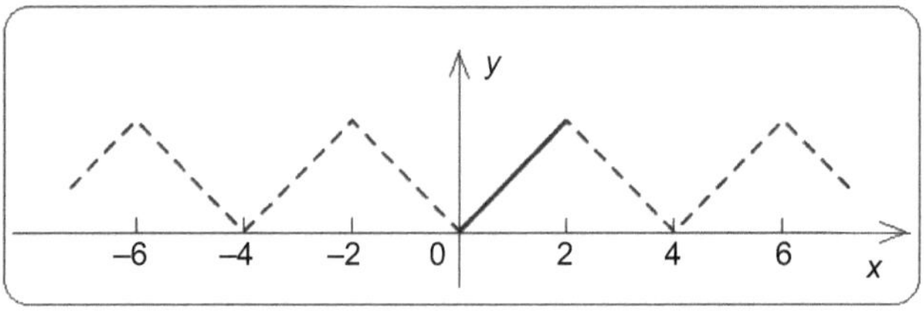


**Figure 6.2** The previous function.

If $n = 0$, then $a_0 = 2$.

Therefore,

$$f(x)=1+\sum_{n=1}^{\infty}\frac{4}{n^2\pi^2}(\cos(n\pi)-1)\cos\frac{n\pi x}{2}.$$

Now, we write the Parseval's identity, first computing the average of $[f(x)]^2$. It is:

$$\frac{1}{2}\int_{-2}^{2}(f(x))^2dx=\frac{1}{2}\int_{-2}^{2}x^2dx=\frac{8}{3},$$

then the average is:

$$\frac{a_0^{\ 2}}{2}+\sum_{n=1}^{\infty}\left(a_n^{\ 2}+b_n^{\ 2}\right)=\frac{2^2}{2}+\sum_{n=1}^{\infty}\frac{16}{n^4\pi^4}(\cos(n\pi)-1)^2.$$

The Parseval's identity becomes:

$$\frac{8}{3}=2+\frac{64}{\pi^4}\left(\frac{1}{1^4}+\frac{1}{3^4}+\frac{1}{5^4}+.....\right)$$

or $$\left(\frac{1}{1^4}+\frac{1}{3^4}+\frac{1}{5^4}+.....\right)=\frac{\pi^4}{96}.$$

This shows that it is possible to use the Parseval's identity for finding the sum of an infinite series.

## 6.7 Integration and Differentiation of a Fourier Series

The Fourier series of a function $f(x)$ may always be integrated term-by-term to give a new series which converges to the integral of $f(x)$. If $f(x)$ is a continuous function of $x$ for all $x$ and is periodic (with period $2\pi$) outside the interval $-\pi < x < \pi$, then term-by-term differentiation of the Fourier series of $f(x)$ leads to the Fourier series of $f'(x)$, provided $f'(x)$ satisfies the Dirichlet conditions. There are numerous applications of Fourier series to solutions of boundary value problems.

*Example*: The RLC circuit.

We consider an RLC circuit driven by a variable voltage $E(t)$ which is periodic, but not necessarily sinusoidal (Fig. 6.3).

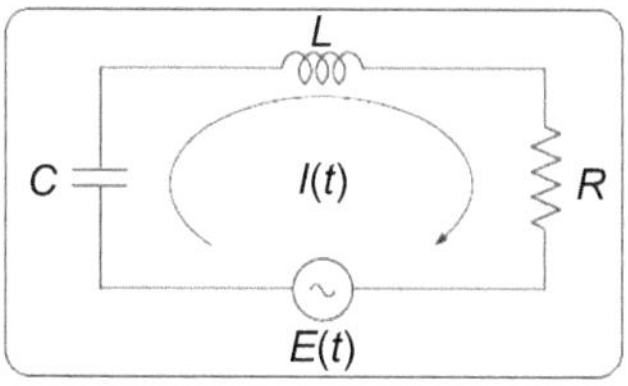


**Figure 6.3** The *RLC* circuit.

We search the current $I(t)$ flowing in the circuit at time $t$. According to Kirchhoff's second law for circuits, the impressed voltage $E(t)$ satisfies the following equation:

$$L\frac{dI}{dt}+RI+\frac{Q}{C}=E(t),$$

with $Q$ the total charge in the capacitor $C$. Being $I = dQ/dt$, through differentiation of the above differential equation, we obtain:

$$L\frac{d^2I}{dt^2}+R\frac{dI}{dt}+\frac{1}{C}I=\frac{dE}{dt}.$$

Under steady-state conditions, the current $I(t)$ is also periodic, with the same period $P$ as for $E(t)$. We assume that both $E(t)$ and $I(t)$ possess Fourier expansions and write them in their complex forms:

$$E(t)=\sum_{n=-\infty}^{+\infty}E_n e^{in\omega t};\ I(t)=\sum_{n=-\infty}^{+\infty}c_n e^{in\omega t}\left(\omega=\frac{2\pi}{P}\right).$$

Assuming that the series can be differentiated term-by-term, we have:

$$\frac{dE}{dt} = \sum_{n=-\infty}^{+\infty} in\omega E_n e^{in\omega t}; \frac{dI}{dt} = \sum_{n=-\infty}^{+\infty} in\omega c_n e^{in\omega t}$$

$$\text{or} \quad \frac{d^2 I}{dt^2} = \sum_{n=-\infty}^{+\infty} (-n^2\omega^2) c_n e^{in\omega t}$$

Substituting these information into the last second-order differential equation, equating the coefficients with the same exponential and solving for $c_n$, we obtain:

$$c_n = \frac{in\omega/L}{\left[(1/LC)^2 - n^2\omega^2\right] + i(R/L)n\omega} E_n ,$$

where $1/LC$ is the 'natural frequency' of the circuit and $R/L$ is the 'attenuation factor'. The Fourier coefficients for $E(t)$ are given by:

$$E_n = \frac{1}{P} \int_{-P/2}^{P/2} E(t) e^{-in\omega t} dt .$$

## 6.8 Multiple Fourier Series

A Fourier expansion of a function of two or three variables is often very useful in many applications. Let us consider the case of a function of two variables $f(x,y)$. For example, we can expand $f(x,y)$ into a double Fourier sine series:

$$f(x,y) = \sum_{m=1}^{\infty} \sum_{n=1}^{\infty} B_{mn} \sin\frac{m\pi x}{L_1} \sin\frac{n\pi y}{L_2} , \tag{6.17}$$

with

$$B_{mn} = \frac{4}{L_1 L_2} \int_0^{L_1} \int_0^{L_2} f(x,y) \sin\frac{m\pi x}{L_1} \sin\frac{n\pi y}{L_2} dxdy . \tag{6.18}$$

Similar expansions can be considered for cosine series and for series having both sines and cosines. To obtain the coefficients $B_{mn}$, we can rewrite $f(x,y)$ as:

$$f(x,y) = \sum_{m=1}^{\infty} C_m \sin\frac{m\pi x}{L_1} , \tag{6.19}$$

with

$$C_m = \sum_{n=1}^{\infty} B_{mn} \sin\frac{n\pi y}{L_2}. \tag{6.20}$$

These ideas can be generalised to triple Fourier series, etc. and are very useful in solving, for example, wave propagation and heat conduction problems in two or three dimensions.

## 6.9 Fourier Integrals and Fourier Transforms

The properties of Fourier series allow the expansion of any periodic function that satisfies the Dirichlet conditions. Many problems in science do not involve periodic functions, therefore, it is desirable to generalise the Fourier series method for including non-periodic functions. We can consider a non-periodic function as a limit of a given periodic function whose period becomes infinite.

***Example* 6.2:** Let us consider the Gaussian probability function:

$$f(x) = Ne^{-\alpha x^2},$$

with $N$ and $\alpha$ constant. Determine:

(a) its Fourier transform $g(\omega)$;
(b) graph $f(x)$;
(c) graph $g(\omega)$.

***Solution*:**

(a) The Fourier transform is given by:

$$g(\omega) = \frac{1}{\sqrt{2\pi}} \int_{-\infty}^{+\infty} f(x) e^{-i\omega x} dx = \frac{N}{\sqrt{2\pi}} \int_{-\infty}^{+\infty} e^{-\alpha x^2} e^{-i\omega x} dx.$$

We can simplify the integral with a change of variable. We note that:

$$-\alpha x^2 - i\omega x = -(x\sqrt{\alpha} + i\omega/2\sqrt{\alpha})^2 - \omega^2/4\alpha.$$

Writing:

$$u = x\sqrt{\alpha} + i\omega/2\sqrt{\alpha},$$

we obtain:

$$g(\omega)=\frac{N}{\sqrt{2\pi\alpha}}e^{-\omega^2/4\alpha}\int_{-\infty}^{+\infty}e^{-u^2}du=\frac{N}{\sqrt{2\alpha}}e^{-\omega^2/4\alpha}.$$

(b and c) In relation to graphs, we have (Figs. 6.4 and 6.5):

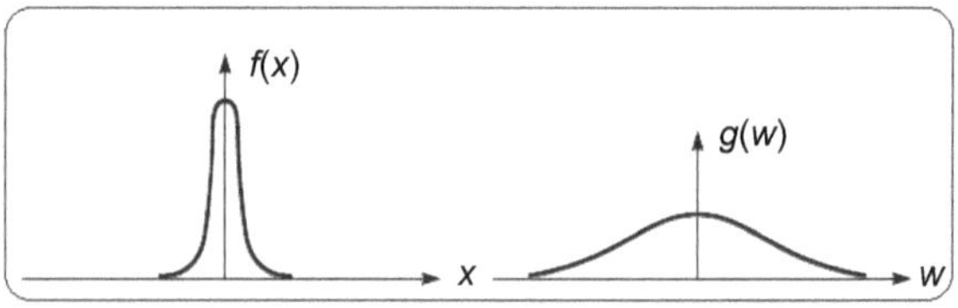


**Figure 6.4** Graph $f(x)$.

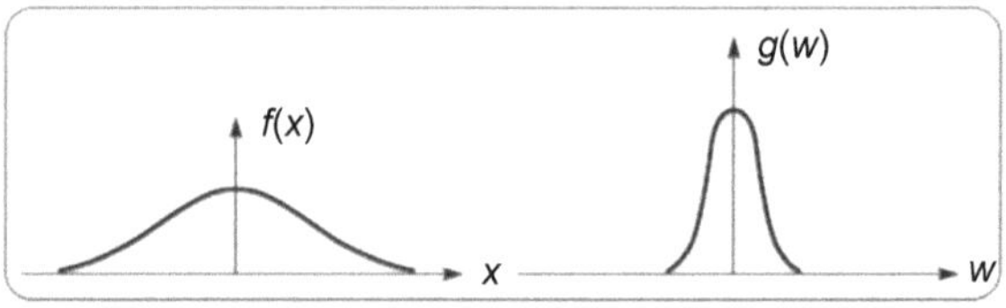


**Figure 6.5** Graph $g(\omega)$.

$g(\omega)$ is a Gaussian probability function too, with a peak at the origin; it monotonically decreases as $\omega \to \pm\infty$. For large $\alpha$, $f(x)$ is sharply peaked and $g(\omega)$ is flattened and vice versa; this is a general feature of Fourier transforms.

## 6.10 Fourier Transforms for Functions of Several Variables

It is possible to extend the development of Fourier transforms to a function of several variables, such as $f(x,y,z)$. Decomposing the function into a Fourier integral with respect to $x$, we have:

$$f(x,y,z)=\frac{1}{\sqrt{2\pi}}\int_{-\infty}^{+\infty}\gamma(\omega_x,y,z)e^{i\omega_x x}d\omega_x, \tag{6.21}$$

where $\gamma$ is the Fourier transform. In a similar way, we can decompose the function with respect to $y$ and $z$, obtaining:

$$f(x,y,z)=\frac{1}{(2\pi)^{3/2}}\int_{-\infty}^{+\infty} g(\omega_x,\omega_y,\omega_z)e^{i(\omega_x x+\omega_y y+\omega_z z)}d\omega_x d\omega_y d\omega_z \,. \tag{6.22}$$

Considering $\omega_x$, $\omega_y$, $\omega_z$ as the components of a vector $\boldsymbol{\omega}$, we can write Eq. 6.22 as:

$$f(\mathbf{r})=\frac{1}{(2\pi)^{3/2}}\int_{-\infty}^{+\infty} g(\boldsymbol{\omega})e^{i\boldsymbol{\omega}\cdot\mathbf{r}}\, d\boldsymbol{\omega}\,. \tag{6.23}$$

# Chapter 7

# Functions of One Complex Variable

## 7.1 Complex Numbers

The theory of functions of a complex variable is a basic part of mathematical analysis. It provides some of the very useful mathematical tools for science.

The known number system is a result of gradual development. The natural numbers (positive integers 1, 2, ...) were first used in counting. Negative integers and zero (0, −1, −2, ...) then arose to allow solutions of equations such as $x + 3 = 2$. In order to solve equations such as $bx = a$ for all integers $a$ and $b$, with $b$ different from zero, rational numbers (or fractions) were introduced. Irrational numbers are numbers which cannot be expressed as $a/b$, with $a$ and $b$ integers and $b$ different from zero, such as $\sqrt{2}$, $\pi$, $e$.

Rational and irrational numbers are all *real numbers*. However, the real number system is still incomplete. For example, there is no real number $x$ which satisfies the algebraic equation $x^2 + 1 = 0$; there is no real number whose square is equal to −1. Euler introduced the symbol $i = \sqrt{-1}$. Today $i$ is called the *unit imaginary number*. It is postulated that $i$ will behave like a real number in all manipulations involving addition and multiplication.

We introduce a general complex number, in Cartesian form:

$$z = x + iy \tag{7.1}$$

*Mathematics and Physics for Nanotechnology: Technical Tools and Modelling*
Paolo Di Sia

ISBN 978-981-4800-02-0 (Hardcover), 978-0-429-02775-8 (eBook)
www.panstanford.com

and refer to $x$ and $y$ as its real and imaginary parts, denoting them by the symbols 'Re $z$' and 'Im $z$', respectively. A number with $x = 0$ is called a *pure imaginary number.*

The *complex conjugate,* or briefly conjugate, of the complex number $z$ is:

$$z^* = x - iy \tag{7.2}$$

and is called *z-star*. Sometimes we write it $\bar{z}$ and call it *z-bar*.

Complex conjugation can be viewed as replacing $i$ by $-i$ within the complex number.

## 7.2 Basic Operations with Complex Numbers

Two complex numbers $z_1 = x_1 + iy_1$ and $z_2 = x_2 + iy_2$ are 'equal' if and only if $x_1 = x_2$ and $y_1 = y_2$.

In performing operations with complex numbers we can proceed as in the algebra of real numbers, replacing $i^2$ with '$-1$' when it occurs.

Given two complex numbers $z_1 = x_1 + iy_1$ and $z_2 = x_2 + iy_2$, the basic operations we can do are:

(a) Addition:

$$z_1 + z_2 = (x_1 + x_2) + i(y_1 + y_2) \tag{7.3}$$

(b) Subtraction:

$$z_1 - z_2 = (x_1 - x_2) + i(y_1 - y_2) \tag{7.4}$$

(c) Multiplication:

$$z_1 z_2 = (x_1 x_2 - y_1 y_2) + i(x_1 y_2 + y_1 x_2) \tag{7.5}$$

(d) Division:

$$\frac{z_1}{z_2} = \frac{(x_1 x_2 + y_1 y_2)}{(x_2^2 + y_2^2)} + i\frac{(y_1 x_2 - x_1 y_2)}{(x_2^2 + y_2^2)} \tag{7.6}$$

## 7.3 Polar Form of a Complex Number

All real numbers can be visualised as points on a straight line (an axis). A complex number, containing two real numbers, can be represented by a point in a two-dimensional plane, known as the *z-plane*, or *complex plane*, or *Gauss plane* (Fig. 7.1).

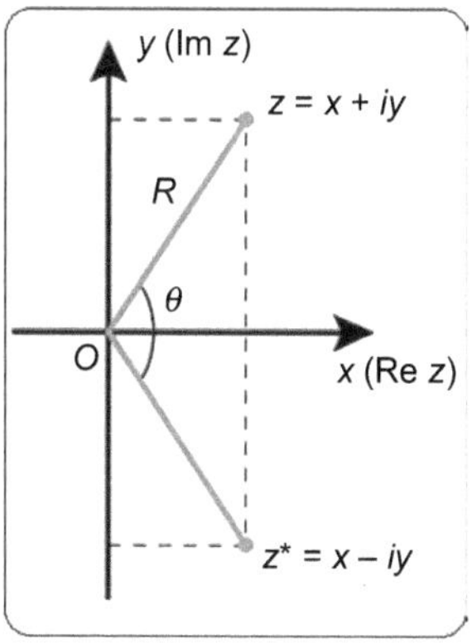


**Figure 7.1** The complex plane.

The complex variable can also be represented by the plane polar coordinates $(r, \theta)$:

$$z = r(\cos\theta + i\sin\theta). \tag{7.7}$$

Using the Euler's formula, we have:

$$e^{i\theta} = r(\cos\theta + i\sin\theta), \tag{7.8}$$

it is possible to rewrite Eq. 7.7 in polar form:

$$z = re^{i\theta}; r = \sqrt{x^2 + y^2}, \tag{7.9}$$

where $r$ is the modulus or absolute value of $z$ and denoted by $|z|$ or 'mod $z$' and $\theta$ is the phase or argument of $z$ and denoted by 'arg $z$'.

If $z_1, z_2, \ldots, z_n$ are complex numbers, the following properties hold for the absolute value of $z$:

(a) $$|z_1 z_2 \cdots z_n| = |z_1||z_2|\cdots|z_n| \tag{7.10}$$

(b) $$\left|\frac{z_1}{z_2}\right| = \frac{|z_1|}{|z_2|}, z_2 \neq 0 \tag{7.11}$$

(c) $$|z_1 + z_2 + \ldots + z_n| \leq |z_1| + |z_2| + \ldots + |z_n| \tag{7.12}$$

(d) $$|z_1 \pm z_2| \geq |z_1| - |z_2| \tag{7.13}$$

***Example* 7.1:** Find $(1 + i)^6$.

***Solution*:** Starting by $z = 1 + i$, we have:

$$z = 1 + i = r(\cos\theta + i\sin\theta) = \sqrt{2}\, e^{i\pi/4}.$$

Therefore, it is: $(1 + i)^6 = (\sqrt{2}\, e^{i\pi/4})^6 = 8\, e^{i3\pi/2}$.

***Example* 7.2:** Calculate in polar form $[(1 + i\sqrt{3}) / (1 - i\sqrt{3})]^{10}$.

***Solution*:** We have:

$$[(1 + i\sqrt{3}) / (1 - i\sqrt{3})]^{10} = (2\, e^{i\pi/3} / 2\, e^{-i\pi/3})^{10} = e^{i20\pi/3}.$$

## 7.4 De Moivre's Theorem and Roots of Complex Numbers

Considering a product of $n$ equal complex numbers, we get:

$$(\cos\theta + i\sin\theta)^n = \cos(n\theta) + i\sin(n\theta). \tag{7.14}$$

This result leads to the 'de Moivre's theorem', a general rule for calculating the $n$-th power of a complex number $z$:

$$z^n = r^n(\cos\theta + i\sin\theta)^n = r^n[\cos(n\theta) + i\sin(n\theta)]. \tag{7.15}$$

From Eq. 7.15, we can get a rule for calculating the $n$-th root of a complex number; if $z = w^n$, then $w = z^{1/n}$ and considering $z$ in Eq. 7.7, we have:

$$w_0 = \sqrt[n]{r}\left(\cos\frac{\theta}{n} + i\sin\frac{\theta}{n}\right). \tag{7.16}$$

This is the $n$-th root of $z$ because $w_0^{\,n} = z$. Considering the periodicity of 'sin' and 'cos' functions, the general rule for calculating the $n$-th root of a complex number is:

$$w = \sqrt[n]{r}\left(\cos\frac{\theta + 2k\pi}{n} + i\sin\frac{\theta + 2k\pi}{n}\right), \tag{7.17}$$

with $k = 0, 1, 2, \ldots, n - 1$. Normally the number corresponding to $k = 0$ (i.e., $w_0$) is said the *principal root* of z. The $n$-th roots of a complex number $z$ are located at the vertices of a regular polygon of $n$ sides inscribed in a circle of radius $\sqrt[n]{r}$.

*Example*: Let us consider the cube roots of 8. As a complex number, 8 is $z = r(\cos\theta + i\sin\theta) = 8 + i\,0$. So, $r = 2$ and the principal argument is $\theta = 0$. Relation of Eq. 7.17 becomes (Fig. 7.2):

$$\sqrt[3]{8} = 2\left(\cos\frac{2k\pi}{3} + i\sin\frac{2k\pi}{3}\right), \text{ with } k = 0, 1, 2.$$

The solution '2' corresponds to $k = 0$, $\theta = 0^0$, '$-1 + i\sqrt{3}$' to $k = 1$, $\theta = 120^0$, '$-1 - i\sqrt{3}$' to $k = 1$, $\theta = 240^0$.

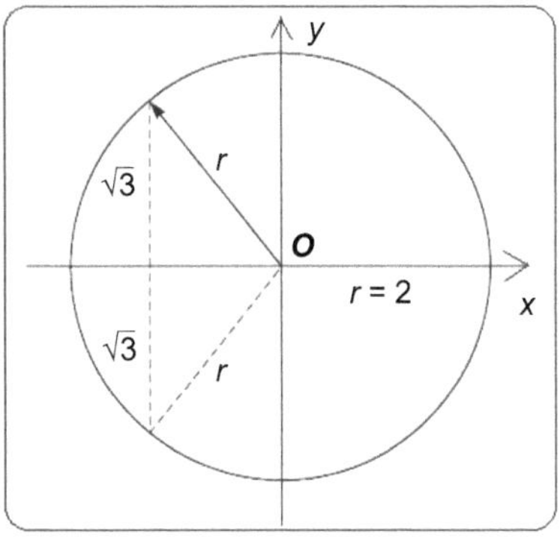


**Figure 7.2** Cube roots of 8.

## 7.5 Functions of a Complex Variable

If to each value of a complex variable $z$, it corresponds one or more values of a complex variable $w$, we say that $w$ is a *function* of $z$ and write $w = f(z)$. The variable $z$ is also called *independent variable* and $w$ is *dependent variable*. If only one value of $w$ corresponds to each value of $z$, we say that $w$ is a 'single-valued' function of $z$ and if more than one value of $w$ corresponds to each value of $z$, $w$ is a 'multiple-valued' function of $z$. We can write:

$$w = u + iv = f(x + iy), \tag{7.18}$$

with $u$ and $v$ real numbers.

*Example*: Let us consider the function $w = z^2$. This function is single-valued, but not one-to-one. It is a two-to-one mapping, since $z$ and $-z$ give the same square. For example, $z_1 = -2 + i$ and $(-z_1) = 2 - i$ are mapped into the same point $w = 3 - 4i$.

## 7.6 Limits and Continuity

We say that $f(z)$ has 'limit' $w_0$ as $z$ approaches $z_0$ and write:

$$\lim_{z \to z_0} f(z) = w_0, \quad \text{if:} \tag{7.19}$$

(a) $f(z)$ is defined and single-valued in a neighbourhood of $z = z_0$, with the possible exception of the point $z_0$ itself;

(b) given any positive number $\varepsilon$ (however small), there exists a positive number $\delta$ such that $|f(z) - w_0| < \varepsilon$ whenever $0 < |z - z_0| < \delta$.

The limit is independent of the way in which $z$ approaches $z_0$.

$f(z)$ is 'continuous' at $z = z_0$ if three conditions are met:

(a) $\lim_{z \to z_0} f(z) = w_0$ must exist;

(b) $f(z)$ is defined at $z_0$, i.e., $f(z)_{z=z0} = f(z_0)$;

(c) $w_0 = f(z_0)$.

A function $f(z)$ is said to be continuous in a region $D$ of the $z$-plane if it is continuous at all points of $D$.

Complex polynomials are continuous everywhere. Quotients of polynomials are continuous whenever the denominator does not vanish.

$f(z)$ is said to be 'discontinuous' at a point $P$ if it is not continuous.

## 7.7 Derivatives and Analytic Functions

Given a continuous, single-valued function of a complex variable $f(z)$ in some region $D$ of the $z$-plane, the derivative $f'(z)$ at some fixed point $z_0$ in $D$ is defined as:

$$f'(z_0) = \frac{df}{dz}(z_0) = \lim_{z \to z_0} \frac{f(z) - f(z_0)}{z - z_0}, \tag{7.20}$$

with $z$ any point of some neighbourhood of $z_0$.

If $f'(z)$ exists at $z_0$ and at every point $z$ in some neighbourhood of $z_0$, then $f(z)$ is said to be 'analytic' at $z_0$.

$f(z)$ is said 'analytic in a region' $D$ of the complex $z$-plane if it is analytic at every point in $D$.

For complex functions, rules for differentiating sums, products and quotients are, in general, the same as for real-valued functions. So, if $f'(z_0)$ and $g'(z_0)$ exist, we have:

$$\text{(a)}\quad (f \pm g)'(z_0) = f'(z_0) \pm g'(z_0) \tag{7.21}$$

$$\text{(b)}\quad (fg)'(z_0) = f'(z_0)g(z_0) + f(z_0)g'(z_0) \tag{7.22}$$

$$\text{(c)}\quad \left(\frac{f}{g}\right)'(z_0) = \frac{f'(z_0)g(z_0) - f(z_0)g'(z_0)}{(g(z_0))^2}, g(z_0) \neq 0 \tag{7.23}$$

## 7.8 Cauchy–Riemann Conditions

The Cauchy–Riemann conditions for the analyticity of $f(z) = u(x,y) + i\, v(x,y)$ are as follows:

$$\frac{\partial u}{\partial x} = \frac{\partial v}{\partial y} \text{ and } \frac{\partial u}{\partial y} = -\frac{\partial v}{\partial x}. \tag{7.24}$$

If the function $f(z) = u(x,y) + i\, v(x,y)$ is analytic in a region $D$, then $u(x,y)$ and $v(x,y)$ satisfy the Cauchy–Riemann conditions at all points of $D$.

*Example*: Let us consider the function: $f(z) = z^2 = x^2 - y^2 + 2\, i\, x\, y$. It is:

$f'(z) = 2\, z$;

$$\frac{\partial u}{\partial x} = 2x = \frac{\partial v}{\partial y} \text{ and } \frac{\partial u}{\partial y} = -2y = -\frac{\partial v}{\partial x}.$$

The Cauchy–Riemann conditions (Eq. 7.24) hold at all points $z$. But it is possible to find examples in which $u(x,y)$ and $v(x,y)$ satisfy the Cauchy–Riemann conditions at $z = z_0$, but $f'(z_0)$ does not exist.

The analyticity is guaranteed if we add an additional hypothesis:

Given $f(z) = u(x,y) + i\, v(x,y)$, if $u(x,y)$ and $v(x,y)$ are continuous with continuous first partial derivatives and satisfy the Cauchy–Riemann conditions (7.24) at all points in a region $D$, then $f(z)$ is analytic in $D$.

Analytic functions are also called *regular* or *holomorphic* functions. A function $f(z)$ is said to be 'singular' at $z = z_0$, if it is not differentiable there; the point $z_0$ is called a *singular point* of $f(z)$.

## 7.9 Harmonic Functions

Let us consider $f(z) = u(x,y) + i\, v(x,y)$, analytic in some region of the $z$-plane; at every point of the region the Cauchy–Riemann conditions (Eq. 7.24) are satisfied. Considering the partial derivative of the first one with respect to $x$ and the partial derivative of the second one with respect to $y$, we get:

$$\frac{\partial^2 u}{\partial x^2} = \frac{\partial^2 v}{\partial x \partial y} \text{ and } \frac{\partial^2 u}{\partial y^2} = -\frac{\partial^2 v}{\partial y \partial x}, \tag{7.25}$$

if these second derivatives exist [it is possible to show that if $f(z)$ is analytic in some region $D$, all its derivatives exist and are continuous in $D$]. From Eq. 7.25, we get:

$$\frac{\partial^2 u}{\partial x^2} + \frac{\partial^2 u}{\partial y^2} = 0. \tag{7.26}$$

Now, considering the partial derivative of the first one with respect to $y$ and the partial derivative of the second one with respect to $x$, we get:

$$\frac{\partial^2 v}{\partial x^2} + \frac{\partial^2 v}{\partial y^2} = 0. \tag{7.27}$$

Equations (7.26) and (7.27) are the 'Laplace partial differential equations' in two independent variables $x$ and $y$. Any function that has continuous partial derivatives of second order and that satisfies the Laplace equation is called *harmonic function*.

## 7.10 Singular Points

A *singular point* (or *singularity*) is a point at which $f(z)$ fails to be analytic; the Cauchy–Riemann conditions break down at a singularity. We have:

(a) *Isolated singular point*: The point $z = z_0$ is called an isolated singular point of $f(z)$ if we can find $\delta > 0$ such that the circle $|z - z_0| < \delta$ encloses no singular point other than $z_0$. If not, $z_0$ is a *non-isolated singularity*.

(b) *Poles*: If it is possible to find a positive integer $n$ such that:

$$\lim_{z \to z_0} (z - z_0)^n f(z) = L \neq 0, \qquad (7.28)$$

then $z = z_0$ is called a *pole of order n*. If $n = 1$, $z_0$ is called a *simple pole*.

*Example*: $f(z) = 1/(z - 3)$ has a simple pole at $z = 3$. $f(z) = 1/(z - 3)^2$ has a pole of order 2 at $z = 3$.

(c) *Branch point*: A function has a branch point at $z_0$ if, upon encircling $z_0$ and returning to the starting point, the function does not return to the starting value. The function is a multiple-valued function.

(d) *Removable singularities*: The singular point $z_0$ is called a 'removable singularity' of $f(z)$ if $\lim_{z \to z_0} f(z)$ exists.

*Example*: Given $f(z) = \sin(z)/z$, the singular point at $z = 0$ is a removable singularity, since $\lim_{z \to 0} \sin(z)/z = 1$.

(e) *Essential singularity*: The function has poles of arbitrarily high order, which cannot be eliminated by multiplication with $(z - z_0)^n$ for any finite choice of $n$.

*Example*: The function $f(z) = \exp\,[1 / (z - 3)]$ has an essential singularity at $z = 3$.

(f) *Singularity at infinity*: The singularity of $f(z)$ at $z = \infty$ is the same type as that of $f(1/w)$ at $w = 0$.

*Example*: $f(z) = z^3$ has a pole of order 3 at $z = \infty$, since $f(1/w) = w^{-3}$ has a pole of order 3 at $w = 0$.

## 7.11 Complex Elementary Functions

(a) *Exponential function*: The exponential function is of fundamental importance and also as a basis for defining the other elementary functions. In its definition, the idea is to preserve many of the characteristic properties of the real exponential function $e^x$. In particular:

(i) $e^z$ is single-valued and analytic;

(ii) $de^z/dz = e^z$;

(iii) $e^z$ reduces to $e^x$ when Im $z = 0$.

The previous requirements are met for:

$$e^z = e^{x+iy} = e^x[\cos(y) + i\sin(y)]. \quad (7.29)$$

We adopt Eq. 7.29 as definition of $e^z$. It is analytic at each point in the entire $z$-plane. The right-hand side of Eq. 7.29 is in the standard polar form with the modulus of $e^z$ given by $e^x$ and argument by $y$.

From Eq. 7.29, with $y = 2\pi$, we obtain the Euler's formula: $\exp(2\pi i) = 1$.

$e^z$ is periodic with imaginary period '$2\pi i$':

$$e^z = e^{z \pm 2\pi n i}\,(n = 0, 1, 2, .....). \quad (7.30)$$

Considering that all values of $w = e^z$ are assumed in the strip: $\pi < y \leq \pi$, this infinite strip is called the *fundamental region* of $e^z$.

(b) *Trigonometric functions*: Considering that from the Euler's formula, we get:

$$\cos x = \frac{e^{ix} + e^{-ix}}{2}; \quad \sin x = \frac{e^{ix} - e^{-ix}}{2i} \quad \text{(with } x \text{ real)}, \quad (7.31)$$

we can define for complex $z$:

$$\cos z = \frac{e^{iz} + e^{-iz}}{2}; \quad \sin z = \frac{e^{iz} - e^{-iz}}{2i}. \quad (7.32)$$

The other trigonometric functions are defined in the usual way:

$$\tan z = \frac{\sin z}{\cos z}, \quad (7.33a)$$

$$\cot z = \frac{\cos z}{\sin z}, \quad (7.33b)$$

$$\sec z = \frac{1}{\cos z}, \quad (7.33c)$$

$$\operatorname{cosec} z = \frac{1}{\sin z}, \quad (7.33d)$$

with the denominators different from zero. The following formulas hold:

$$\sin(-z) = -\sin z \tag{7.34}$$

$$\cos(-z) = \cos z \tag{7.35}$$

$$\sin^2 z + \cos^2 z = 1 \tag{7.36}$$

$$\cos(z_1 \pm z_2) = \cos z_1 \cos z_2 \mp \sin z_1 \sin z_2 \tag{7.37}$$

$$\sin(z_1 \pm z_2) = \sin z_1 \cos z_2 \pm \cos z_1 \sin z_2 \tag{7.38}$$

$$\frac{d}{dz}(\sin z) = \cos z \tag{7.39}$$

$$\frac{d}{dz}(\cos z) = -\sin z \tag{7.40}$$

The functions 'sin $z$' and 'cos $z$' are analytic for all $z$, 'tan $z$' and 'sec $z$' except the points where cos $z$ is zero, 'cot $z$' and 'cosec $z$' except the points where sin $z$ is zero.

The functions 'cos $z$' and 'sec $z$' are even, 'sin $z$' and 'cosec $z$' are odd.

(c) *Hyperbolic functions*: Hyperbolic functions can be introduced via imaginary circular angles. Using the definitions of the hyperbolic functions of real variables, we can write:

$$\sin(iy) = i\sinh y \tag{7.41}$$

$$\cos(iy) = \cosh y \tag{7.42}$$

It is:

$$\sin z = \sin(x + iy) = \sin x \cosh y + i \cos x \sinh y \tag{7.43}$$

$$\cos z = \cos(x + iy) = \cos x \cosh y - i \sin x \sinh y \tag{7.44}$$

Taking $x = 0$ in Eqs. 7.43 and 7.44, we get Eqs. 7.41 and 7.42. We note the big difference between complex and real sin or cos functions. The real functions are bounded between −1 and +1 and the complex functions can take on arbitrarily large values.

(d) *Logarithmic function*: We define $w = \ln z$ considering that $e^w = z$ for each $z \neq 0$. Setting $w = u + i\,v$ and $z = r\exp(i\theta) = |z|\exp(i\theta)$, we have:

$$e^w = e^u e^{iv} = re^{i\theta} \tag{7.45}$$

It follows that:

$$e^u = r = |z|; v = \theta = \arg z \tag{7.46}$$

Therefore:

$$w = \ln z = \ln r + i\theta = \ln|z| + i\arg z. \tag{7.47}$$

Considering that the argument of $z$ is determined in multiples of $2\pi$, the complex natural logarithm is infinitely many-valued. Then we can rewrite Eq. 7.47 as:

$$\ln z = \ln|z| + i(\theta + 2k\pi); \quad k = 0, \pm 1, \pm 2, \ldots.., \tag{7.48}$$

and $\theta_1$ is the 'principal argument' of $z$, i.e., with $0 \leq \theta < 2\pi$. For any particular value of $k$, the logarithm becomes single-valued; $k = 0$ corresponds to the principal value.

The real logarithm $y = \ln x$ makes sense if $x > 0$; in the complex case, we can have also negative numbers.

*Example*: Let us consider $\ln z = \ln(-3)$.
It is $\ln(-3) = \ln|-3| + i \arg(-3) = \ln 3 + i(\pi + 2k\pi)$.
Its principal value is the complex number $\ln 3 + i\pi$.

## Exercises

7.1 Calculate the power of $i$: $i^{41}$. [$i$]

7.2 Simplify the expression: $(3 + i)(3 - i)(1/5 + 1/10\, i)$. [$2 + i$]

7.3 Verify whether $z = -1 \pm 2i$ satisfies the equation:
$z^3 + z^2 + 3z - 5 = 0$. [Yes]

7.4 Calculate the modulus of $1 + i - i/(1 - 2i)$. [$\frac{1}{5}\sqrt{65}$]

7.5 Write the exponential and trigonometric form of:
$z = 1/(3 + 3i)$.
[$z = (\sqrt{2}/6) \exp\{(7/4)\pi i\} = (\sqrt{2}/6)\{\cos((7/4)\pi) + i \sin((7/4)\pi)\}$]

7.6 Calculate the power $z^6$ of $z = (1 + i)/(2 - 2i)$. [$-1/64$]

7.7 Calculate the roots of $\sqrt[3]{1+i}$.

[ $\sqrt[6]{2}\exp(\pi i/12)$ ; $\sqrt[6]{2}\exp(3\pi i/4)$ ; $\sqrt[6]{2}\exp(17\pi i/12)$ ]

7.8 Calculate the cubic roots of $z = \exp(\pi i/6) + \exp(\pi i/2)$ .

[ $\sqrt[6]{3}\exp(\pi i/9)$ ; $\sqrt[6]{3}\exp(7\pi i/9)$ ; $\sqrt[6]{3}\exp(13\pi i/9)$ ]

7.9 Solve the equation: $z^2 + 3iz + 4 = 0$ and represent the solutions in the Gauss plane.

[ $z_1 = -4i$ ; $z_2 = i$ ; (0, −4); (0, 1)]

7.10 Solve the equation: $z\bar{z} - z + i/4 = 0$ and represent the solutions in the Gauss plane.

[ $z_1 = ((2+\sqrt{3})/4) + i/4$ ; $z_2 = ((2-\sqrt{3})/4) + i/4$ ;
$((2+\sqrt{3})/4, 1/4)$ ; $((2-\sqrt{3})/4, 1/4)$ ]

Chapter 8

# Complex Integration

## 8.1 Line Integrals in the Complex Plane

Let us consider a smooth curve $C$ with a finite length in the complex $z$ plane; let $f(z)$ be continuous at all points of $C$ (Fig. 8.1).

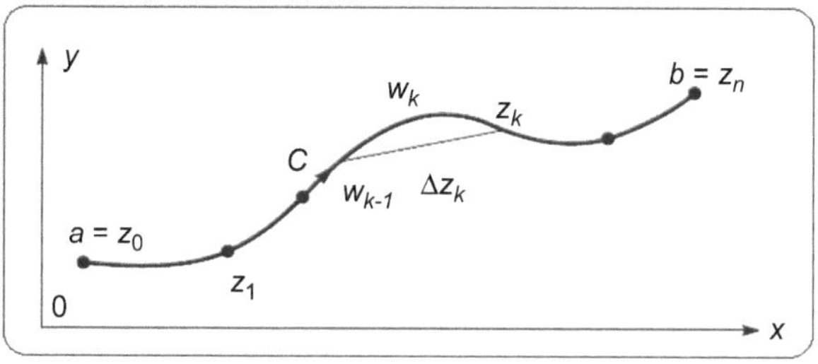


**Figure 8.1** Complex line integral.

Let us sub-divide $C$ into $n$ parts by means of arbitrarily chosen points $a = z_0, z_1, z_2, \ldots.., z_{n-1}, b = z_n$. We choose a point $w_k$ on each arc joining $z_{k-1}$ to $z_k$ ($k = 1, 2, \ldots.., n$) and write the sum:

$$S_n = \sum_{k=1}^{n} f(w_k)\Delta z_k; \quad \Delta z_k = z_k - z_{k-1}. \tag{8.1}$$

Increasing $n$ in such a way that the largest of the chord lengths $|\Delta z_k|$ approaches zero, the sum $S_n$ approaches a limit. If this limit exists

*Mathematics and Physics for Nanotechnology: Technical Tools and Modelling*
Paolo Di Sia

ISBN 978-981-4800-02-0 (Hardcover), 978-0-429-02775-8 (eBook)
www.panstanford.com

and has the same value independently from the choice of $z_j$-s and $w_j$-s, then this limit is called *contour integral* (with contour $C$) or *line integral* of $f(z)$ along $C$ and is denoted by:

$$\int_C f(z)dz \text{ or } \int_a^b f(z)dz. \tag{8.2}$$

If $C$ is closed ($a \equiv b$), we write:

$$\oint f(z)dz. \tag{8.3}$$

Reversing the direction of integration, it changes the sign of the integral. Complex integrals have the following properties:

$$\int_C [f(z)+g(z)]dz = \int_C f(z)dz + \int_C g(z)dz \tag{8.4}$$

$$\int_C kf(z)dz = k\int_C f(z)dz \ (k = \text{real or complex constant}) \tag{8.5}$$

$$\int_a^b f(z)dz = -\int_b^a f(z)dz \tag{8.6}$$

$$\int_a^b f(z)dz = \int_a^p f(z)dz + \int_p^b f(z)dz \ (a < p < b) \tag{8.7}$$

$$\left|\int_C f(z)dz\right| \leq ML \tag{8.8}$$

($M$ = max $|f(z)|$ on $C$; $L$ = length of $C$)

***Example* 8.1:** Evaluate the integral of the function $(z^*)^2$; $C$ is a straight line joining the points $z = 0$ and $z = 1 + 2i$.

***Solution*:** $(z^*)^2 = (x - iy)^2 = x^2 - y^2 - 2xyi$. Then:

$$\int_C [z^*]^2 dz = \int_C (x^2 - y^2 - 2xyi)\, d(x+iy)$$

$$= \int_C [(x^2 - y^2)dx + 2xydy] + i\int_C [-2xydx + (x^2 - y^2)dy]$$

But, being the Cartesian equation of $C$ equal to $y = 2x$, the integral becomes:

$$\int_C [z^*]^2 dz = \int_0^1 5x^2 dx - i\int_0^1 10x^2 dx = 5/3 - i10/3.$$

## 8.2 Cauchy's Integral Theorem

This theorem states that if $f(z)$ is analytic in a simply-connected region (i.e., if any simple closed curve which lies in it can be shrunk to a point without leaving it) and on its boundary $C$, then it holds:

$$\oint_C f(z)dz = \oint_C (udx - vdy) + i\oint_C (vdx + udy) = 0. \tag{8.9}$$

A simply-connected region has no hole in it (Fig. 8.2).

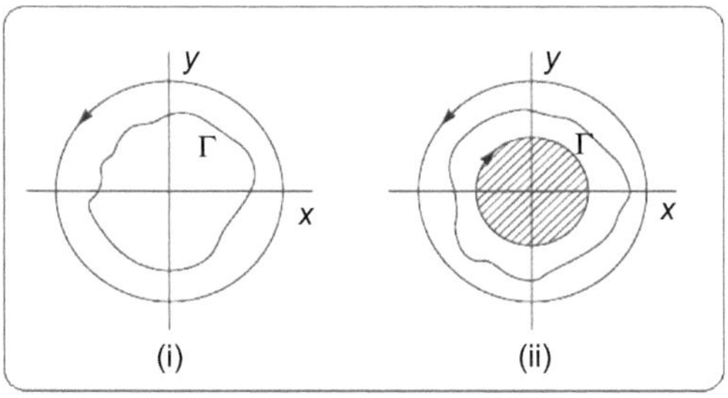


**Figure 8.2** (a) Simply-connected, and (b) multiply-connected region.

Cauchy's theorem is also valid for multiply-connected regions. Considering for simplicity a doubly-connected region, we can follow a path as shown in Fig. 8.3, obtaining the simply-connected region.

An observer walking on the boundary has the domain always on his left. Then Eq. 8.9 becomes:

$$\oint_C f(z)dz = \oint_{C_1} f(z)dz + \oint_{C_2} f(z)dz = 0. \tag{8.10}$$

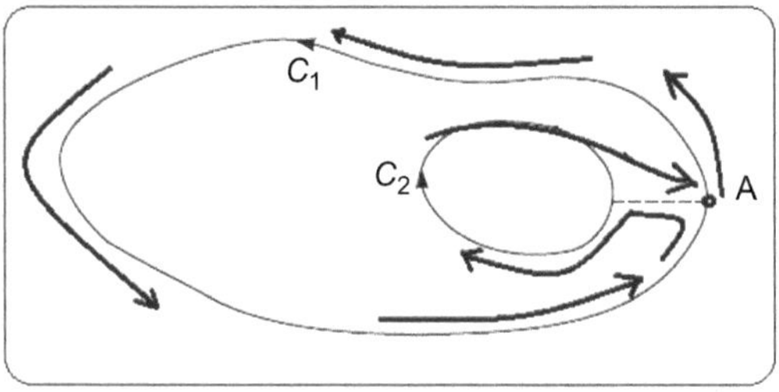


**Figure 8.3** Followed path for avoiding the hole.

An integration contour cannot be moved across a hole or singularity, but it can be made to collapse around it (Fig. 8.4).

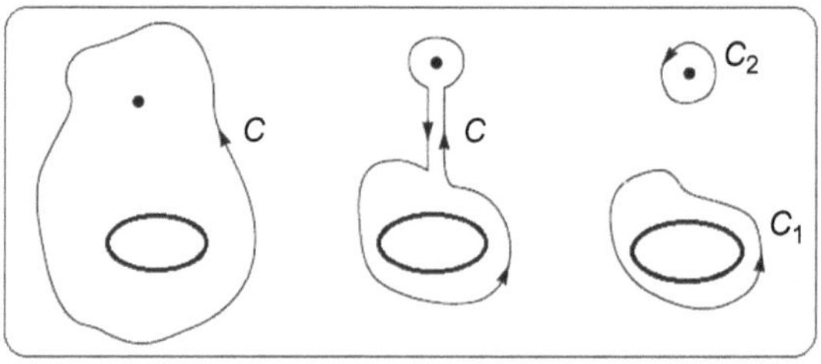


**Figure 8.4** Example of collapsing contour.

For multiply-connected regions, Eq. 8.10 can then be generalised as:

$$\oint_C f(z)dz = \sum_{k=1}^{n} \oint_{C_k} f(z)dz\,. \tag{8.11}$$

## 8.3 Cauchy's Integral Formula

The Cauchy's integral formula is one of the most important consequences of Cauchy's integral theorem. It states that if $f(z)$ is analytic in a simply-connected region $R$ and $z_0$ is any point in the interior of $R$ which is enclosed by a simple closed curve $C$, taking the integration around $C$ in the positive sense (counter-clockwise), it holds (Fig. 8.5):

$$f(z_0) = \frac{1}{2\pi i} \oint_C \frac{f(z)}{z - z_0} dz\,. \tag{8.12}$$

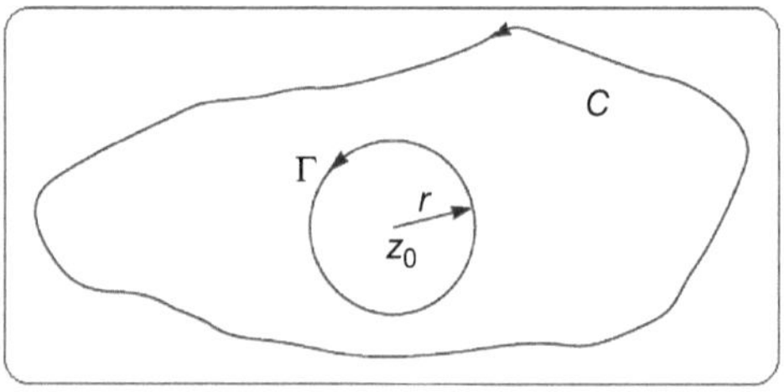


**Figure 8.5** Cauchy's integral formula.

The formula is also true for multiply connected regions.

## 8.4 Series Representations of Analytic Functions

Most of the definitions and theorems related to an infinite series of real terms can be applied with little changes (or no changes) to series with complex terms.

We call 'complex sequence', an ordered list assigning to each positive integer $n$, a complex number $z_n$: $z_1, z_2, \dots, z_n, \dots$

*Example*: $i, i^2, \dots, i^n, \dots$ is a complex sequence.

A sequence is said to be 'convergent' with the limit $l$ if, given $\varepsilon > 0$, we can find a positive integer $N$ such that $|z_n - l| < \varepsilon$ for each $n \geq N$. We write:

$$\lim_{n\to\infty} z_n = l. \tag{8.13}$$

## 8.5 Integration with the Residue Method

The integration with the residue method is useful in evaluating both real and complex integrals. This method is based on the residue theorem.

Let us consider a simple closed curve $C$ containing a number of isolated singularities of a function $f(z)$. Considering around each singular point a small circle enclosing no other singular points, these circles together with the curve $C$ form the boundary of a multiply-connected region in which $f(z)$ is everywhere analytic and to which we can apply the Cauchy's theorem (Fig. 8.6).

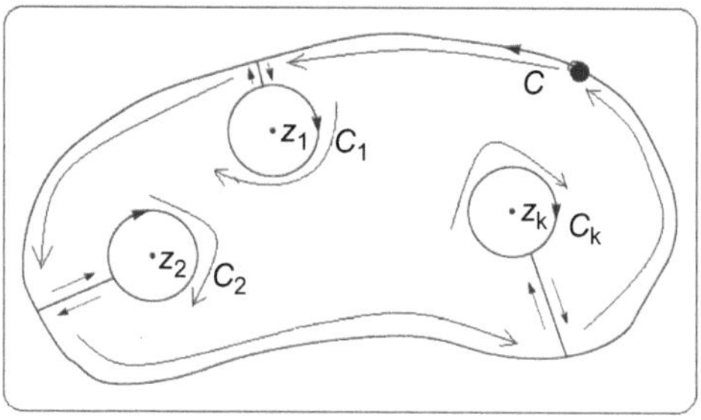


**Figure 8.6** The residue theorem.

The residue theorem states:

If $f(z)$ is an analytic function inside and on a simple closed curve $C$, except at a finite number of singular points $z_1, z_2, \ldots, z_m$ inside $C$, then it holds:

$$\oint_C f(z)dz = 2\pi i \sum_{k=1}^{m} \mathrm{Res}_{z=z_k} f(z) = 2\pi i\left(r_1 + r_2 + \ldots + r_m\right), \quad (8.14)$$

with $r_k$ the residue of $f(z)$ at the singular point $z_k$.

The general formula for getting residues for poles of order $m$ is given by:

$$\mathrm{Res}_{z=a} f(z) = \frac{1}{(m-1)!}\lim_{z\to a}\left(\frac{d^{m-1}}{dz^{m-1}}\left((z-a)^m f(z)\right)\right). \quad (8.15)$$

***Example***: Let us consider the function:

$$f(z) = \frac{-3z+4}{z^2 - z}.$$

Find:

(a) the poles;

(b) the residues;

(c) apply the residue theorem.

***Solution***:

(a) The poles correspond to the values vanishing the denominator: $z^2 - z$, i.e., $z_1 = 0$ and $z_2 = 1$. These are simple poles.

(b) The residues are (Eq. 8.15):

$$\mathrm{Res}_{z=0}\left(\frac{-3z+4}{z^2 - z}\right) = \lim_{z\to 0}\left(z\frac{-3z+4}{z(z-1)}\right) = -4;$$

$$\mathrm{Res}_{z=1}\left(\frac{-3z+4}{z^2 - z}\right) = \lim_{z\to 1}\left(z(z-1)\frac{-3z+4}{z(z-1)}\right) = 1.$$

(c) Using Eq. 8.14, we have:

$$\oint_C \frac{-3z+4}{z^2 - z}dz = 2\pi i(-4+1) = -6\pi i,$$

for every simple closed curve $C$ enclosing the points $z_1 = 0$ and $z_2 = 1$.

If $z_1 = 0$ lies inside $C$ and $z_2 = 1$ outside, we have:

$$\oint_C \frac{-3z+4}{z^2-z}dz = 2\pi i(-4) = -8\pi i \,.$$

The integrations are taken in the counter-clockwise sense.

## 8.6 Evaluation of Real Definite Integrals

The residue theorem gives a useful method for evaluating some classes of complicated real definite integrals. The contour must be closed for applying the residue theorem, whereas many integrals of practical interest involve integration over open curves. Therefore, we must close the paths of integration before applying the residue theorem.

So, the ability to evaluate such integrals strongly depends on how the contour is closed, since it requires knowledge of the additional contributions from the added parts of the closed contour. A number of techniques are known for closing open contours. We limit our discussion to two cases:

(a) ***Improper integrals of a rational function***:

$$\int_{-\infty}^{+\infty} f(x)dx = \lim_{a\to-\infty}\int_a^0 f(x)dx + \lim_{b\to+\infty}\int_0^b f(x)dx \,. \tag{8.16}$$

If both limits exist, it is possible to write:

$$\int_{-\infty}^{+\infty} f(x)dx = \lim_{r\to\infty}\int_{-r}^{r} f(x)dx \,. \tag{8.17}$$

*Assumptions*: $f(x)$ is a real rational function and its denominator is different from zero for all real $x$. The denominator has a degree at least two units higher than that of the numerator. In this case, we can use Eq. 8.17. We consider the corresponding contour integral, with $C$ consisting of the dashed line $\Gamma$ as in Fig. 8.7.

We obtain:

$$\int_{-\infty}^{+\infty} f(x)dx = 2\pi i\sum \text{Res}\, f(z) \,. \tag{8.18}$$

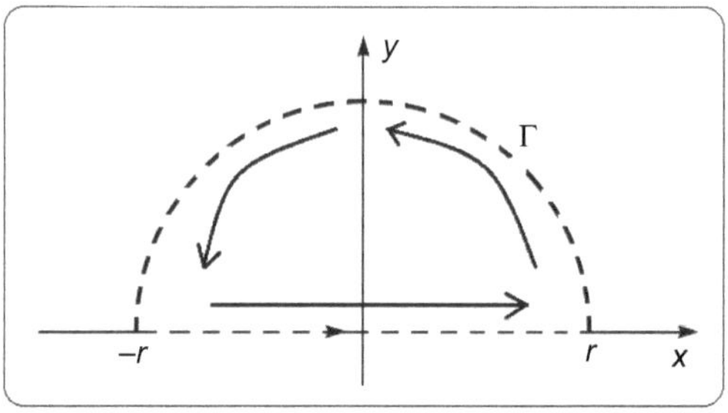


**Figure 8.7** Path Γ of the contour integral.

(b) ***Integrals of rational functions of sinθ and cosθ:***

$$\int_0^{2\pi} f(\sin\theta,\cos\theta)d\theta\,, \tag{8.19}$$

$f$ is a real rational function, finite on the interval $0 \le \theta \le 2\pi$.

Let $z = \exp(i\theta) \to dz = i\exp(i\theta)\,d\theta \to d\theta = dz/iz$. Furthermore, it is: $\sin\theta = (z - z^{-1})/2i$ and $\cos\theta = (z + z^{-1})/2$.

In this way, the integrand becomes a rational function of $z$, say $g(z)$. Considering the interval for $\theta$, the variable $z$ ranges once around the unit circle $|z| = 1$ and the integral becomes:

$$\int_0^{2\pi} f(\sin\theta,\cos\theta)d\theta = \oint_C \frac{g(z)}{iz}dz\,, \tag{8.20}$$

with integration taken in the counter-clockwise sense.

## Exercises

8.1 Determine poles and residues of the function: $1/(1+z^4)$ inside the half-plane $\text{Im}\, z > 0$.

$[\,z_1 = \exp(\pi i/4)\,;\; z_2 = \exp(3\pi i/4)\,;$

$\text{Res}(f,z_1) = 1/4\exp(-3\pi i/4)\,;\; \text{Res}(f,z_2) = 1/4\exp(-\pi i/4)\,]$

8.2 Calculate with the residue method: $\displaystyle\int_{-\infty}^{+\infty} \frac{(a^2-b^2)x^2dx}{(x^2+a^2)(x^2+b^2)}$, $a \ne b$.

$[(a-b)\pi]$

8.3 Calculate with the residue method: $\oint_C \frac{2z\,dz}{(z^2+1)(2z^2-5z+2)}$,

with $C$ the circumference with centre in the origin and $R=\sqrt{2}$.

$$\left[-\frac{8}{15}\pi i\right]$$

8.4 Calculate with the residue method: $\int_0^{2\pi} \frac{d\theta}{(2+\cos\theta)^2}$.

$$\left[\frac{4}{9}\sqrt{3}\pi\right]$$

8.5 Calculate with the residue method: $\int_{-\infty}^{+\infty} \frac{x^2\,dx}{2(x^2+1)^2(x^2+4)}$.

$$\left[\frac{\pi}{18}\right]$$

Chapter 9

# Partial Differential Equations

## 9.1 Introduction

Given a function $f : \Omega \subset R^N \rightarrow R$, with $N$ sufficiently high natural number, we say 'partial differential equation of order $n$', an equation of the form:

$$f\left(x_1,\ldots\ldots,x_m,u,\frac{\partial u}{\partial x_1},\ldots\ldots,\frac{\partial u}{\partial x_m},\ldots\ldots,\frac{\partial^n u}{\partial x_1^n},\ldots\ldots,\frac{\partial^n u}{\partial x_m^n}\right)=0\,, \tag{9.1}$$

if the function explicitly depends from at least one of the partial derivatives of order $n$ of $u$.

$x_1$, ..., $x_m$ are the independent variables, $u(x_1, \ldots, x_m)$ is the unknown function that we want to satisfy the differential equation.

A function $u = \varphi(x_1, \ldots, x_m)$ that identically satisfy Eq. 9.1 is called *integral* or *solution* of the differential equation.

The totality of the integrals, excluded at most some of particular character, called *singular*, is the *general integral* of the equation.

If the function $f$ is a polynomial of degree $\leq 1$ in $u$ and its derivatives and in fact depends on some partial derivative of $u$, then the equation is said *linear*.

A linear equation of the first order has the following expression:

$$P_1\frac{\partial u}{\partial x_1}+P_2\frac{\partial u}{\partial x_2}+\ldots+P_m\frac{\partial u}{\partial x_m}-Qu=R\,, \tag{9.2}$$

---

*Mathematics and Physics for Nanotechnology: Technical Tools and Modelling*
Paolo Di Sia

ISBN 978-981-4800-02-0 (Hardcover), 978-0-429-02775-8 (eBook)
www.panstanford.com

where $P_i$, $Q$ and $R$ are functions of $(x_1, \ldots, x_m)$ and not all $P_i$ are equal to zero.

## 9.2 Linear Second-Order Partial Differential Equations

The second-order linear partial differential equation in two independent variables has the general form:

$$A_1 \frac{\partial^2 u}{\partial x^2} + A_2 \frac{\partial^2 u}{\partial x \partial y} + A_3 \frac{\partial^2 u}{\partial y^2} + A_4 \frac{\partial u}{\partial x} + A_5 \frac{\partial u}{\partial y} + A_6 u = G, \tag{9.3}$$

where the coefficients $A_i$ may be dependent on variables $x$ and $y$. Equation 9.3 is *non-homogeneous*; if the $G$ function is zero, it is called *homogeneous.*

If $u_1, u_2, \ldots, u_n$ are solutions of a linear homogeneous partial differential equation, then $c_1u_1 + c_2u_2 + \ldots + c_nu_n$ is also a solution, with $c_i$ constants ('superposition principle', not working for non-linear equations).

The *general solution* of a linear non-homogeneous partial differential equation is obtained by adding a *particular solution* of the non-homogeneous equation to the general solution of the homogeneous one.

The homogeneous form of Eq. 9.3 looks like the equation of a general conic:

$$ax^2 + bxy + cy^2 + dx + ey + f = 0. \tag{9.4}$$

We say that Eq. 9.3 is of

$$\begin{aligned} &\text{Elliptic type} &&: \; A_2^2 - 4A_1A_3 < 0 \\ &\text{Hyperbolic type} &&: \; A_2^2 - 4A_1A_3 > 0 \\ &\text{Parabolic type} &&: \; A_2^2 - 4A_1A_3 = 0 \end{aligned}$$

*Examples*:

(a) The 2-D Laplace equation $\frac{\partial^2 u}{\partial x^2} + \frac{\partial^2 u}{\partial y^2} = 0$, having $A_1 = A_3 = 1$ and all other coefficients equal to zero, implies $A_2^2 - 4A_1A_3 < 0$, therefore, it is of elliptic type.

(b) The equation $\frac{\partial^2 u}{\partial x^2} - k\frac{\partial^2 u}{\partial y^2} = 0$, with $k$ real positive constant, having $A_1 = 1, A_3 = -k$ and all other coefficients equal to zero, implies $A_2^2 - 4A_1A_3 > 0$, therefore, it is of hyperbolic type.

## 9.3 Important Second-Order Partial Differential Equations

(a) *Laplace's equation*: this equation has the form:

$$\nabla^2 u = 0. \tag{9.5}$$

*Examples*:

The $u$ function may be

(i) the electrostatic potential in a charge-free region

(ii) the gravitational potential in a region without matter

(iii) the velocity potential for an incompressible fluid without sources or wells

(b) *Poisson's equation*: this equation has the form:

$$\nabla^2 u = \rho(x, y, z). \tag{9.6}$$

*Examples*:

(i) If $u$ represents the electrostatic potential in a region containing charges, $\rho$ is proportional to the electrical charge density.

(ii) Considering the gravitational field, in the case of the gravitational potential, $\rho$ is proportional to the mass density in the region.

(c) *Wave equation*: this equation has the form:

$$\nabla^2 u = \frac{1}{v^2}\frac{\partial^2 u}{\partial t^2}. \tag{9.7}$$

*Examples*:

The $u$ represents

(i) the displacement from equilibrium of a vibrating string

(ii) the longitudinal displacement from the equilibrium of a vibrating beam

(iii) a component of electric field $\boldsymbol{E}$ or magnetic field $\boldsymbol{B}$ of an electromagnetic wave

(d) *Heat conduction equation*: this equation has the form:

$$\frac{\partial T}{\partial t} = \alpha \nabla^2 T \,. \tag{9.8}$$

where $T$ is the temperature of a solid at time $t$ and $\alpha$ is the diffusivity.

*Example*: The equation can also be used as a more general diffusion equation. In this case, $u$ is the concentration of a diffusing substance.

# Chapter 10

# Numerical Methods

## 10.1 Interpolation

One of the most common interpolation techniques is the *polynomial interpolation*. Having a set of measured data $(x_0, y_0)$, $(x_1, y_1)$, ….. , $(x_n, y_n)$, we search a representation of them by a smooth curve of the form $y = f(x)$. For analytical convenience, this curve is usually assumed to be of polynomial type:

$$f(x) = a_0 + a_1 x + a_2 x^2 + \ldots\ldots + a_n x^n . \tag{10.1}$$

We use the measured data for evaluating the coefficients $a_i$ in Eq. 10.1. Therefore, we write a system of $n + 1$ equations to solve for the $n + 1$ coefficients $a_0$, $a_1$, ….. , $a_n$:

$$\begin{cases} y_0 = a_0 + a_1 x_0 + a_2 x_0^2 + \ldots\ldots + a_n x_0^n \\ y_1 = a_0 + a_1 x_1 + a_2 x_1^2 + \ldots\ldots + a_n x_1^n \\ \ldots\ldots\ldots\ldots\ldots\ldots\ldots \\ y_n = a_0 + a_1 x_n + a_2 x_n^2 + \ldots\ldots + a_n x_n^n \end{cases} . \tag{10.2}$$

## 10.2 Solutions of Equations: Graphical Method

If it is not possible for different reasons (inability to get exact solutions, analytical computational difficulty, etc.) to find directly

*Mathematics and Physics for Nanotechnology: Technical Tools and Modelling*
Paolo Di Sia

ISBN 978-981-4800-02-0 (Hardcover), 978-0-429-02775-8 (eBook)
www.panstanford.com

the solutions of the equation $f(x) = 0$, we search for approximate solutions. We can rewrite the previous equation as:

$$g(x) = h(x), \tag{10.3}$$

and then we can get the graph of $y = g(x)$ and $y = h(x)$. The $x$ values of the intersection points of the two curves in the plane give the approximate values of the searched solutions. In particular with transcendental equation, this method is useful.

***Example* 10.1:** Find the solutions of the equation $x^3 - 12.34\,x - 7.75 = 0$.

***Solution*:** We can rewrite it as $x^3 = 12.34\,x + 7.75$ and plot the curves (Fig. 10.1):

$y = x^3$ (red dashed curve) and $y = 12.34\,x + 7.75$ (green solid line).

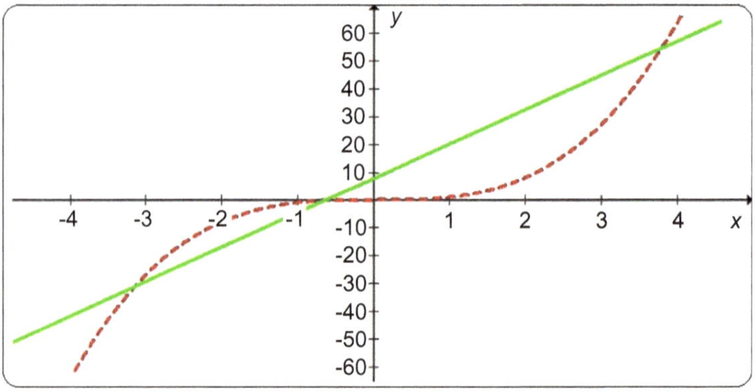


**Figure 10.1** Curves of Example 10.1.

From Fig. 10.1, we get $x_1 = -3.14$; $x_2 = -0.65$; and $x_3 = 3.79$.

***Example* 10.2:** Find the solutions of the equation $e^{3x} + 5.2\cos(3x) = 0$.

***Solution*:** We plot the curves: $y = e^{3x}$ (red dashed curve) and $y = -5.2\cos(3x)$ (green solid line) and search the $x$ values of the intersection points (Fig. 10.2).

From Fig. 10.2, we get for example in the interval $I_1 = (-2, 0)$:

$x_1 = -1.57$; $x_2 = -0.54$.

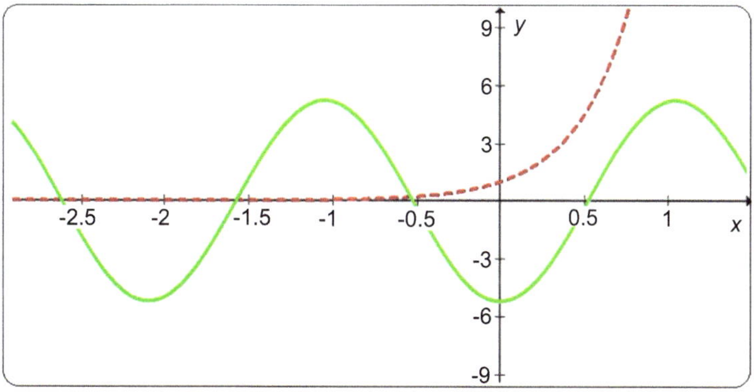


**Figure 10.2** Curves of Example 10.2.

For improving the accuracy of values, it is possible to zoom on particular parts of the graphs.

## 10.3 Method of Linear Interpolation

With this method, also called *method of false position*, we proceed as follows:

If $x_1$ is an approximate value of the solution $x_0$ of the equation $f(x) = 0$, writing $x = x_1$ on the right-hand of Eq. 10.3 and obtain $g(x) = h(x_1)$, supposing that we can solve this one. If the solution is $x_2$, we write $x = x_2$ on the right-hand of Eq. 10.3: $g(x) = h(x_2)$. Iterating the process, we obtain for the $n$th approximation:

$$g(x) = h(x_{n-1}). \tag{10.4}$$

From geometric considerations, the sequence $x_1, x_2, ..., x_n$ converges to the root $x = x_0$ if in the interval: $2|x_1 - x_0|$, the following conditions hold:

(a) $|g'(x)|>|h'(x)|$; (10.5)

(b) the derivatives are bounded.

***Example* 10.3:** Find the approximate values of the real roots of the equation $e^x - 3x = 0$ with an accuracy of three decimals.

We can rewrite the equation as $x = e^x/3$. According to Eq. 10.4, we have $x_{n+1} = e^{x_n}/3$. There are two roots (Fig. 10.3).

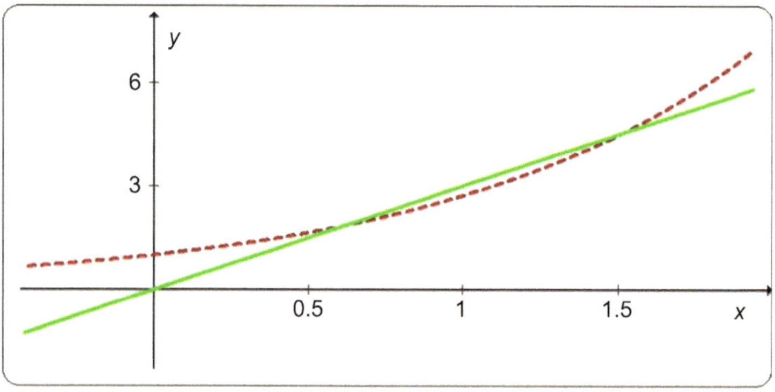


**Figure 10.3** Curves of Example 10.3.

The first root is around $x = 0.6$ and the second around $x = 1.5$. Considering $x_1 = 0.6$, we have:

$$x_2 = e^{x_1}/3 = 0.60626 \ldots$$
$$x_3 = e^{x_2}/3 = 0.61005 \ldots$$
$$x_4 = e^{x_3}/3 = 0.61236 \ldots$$
$$x_5 = e^{x_4}/3 = 0.61377 \ldots$$
$$x_6 = e^{x_5}/3 = 0.61464 \ldots$$
$$x_7 = e^{x_6}/3 = 0.61517 \ldots$$
$$x_8 = e^{x_7}/3 = 0.61550 \ldots$$

We can stop our computation. The found value is 0.615. In the same way, we can proceed with the second root.

## 10.4 Newton Method

With this method, the successive terms in the sequence of approximate values $x_1, x_2, \ldots, x_n$ converging to the root consider the intersection with the $x$-axis of the tangent line to the curve $y = f(x)$. The relation appearing by this method is:

$$x_{n+1} = x_n - \frac{f(x_n)}{f'(x_n)}, \; n = 1, 2, \ldots \qquad (10.6)$$

If the function has a bad behaviour near the root, the method may fail.

***Example* 10.4:** Solve with the Newton's method the equation: $x^3 - 2 = 0$ with a 'four-decimal place' accuracy.

***Solution*:** We note that: $1 < 2^{1/3} < 2$. If we take $x_1 = 1.5$, Eq. 10.6 gives:

$$x_2 = 1.29629629\ldots,$$
$$x_3 = 1.26093222\ldots,$$
$$x_4 = 1.25992186\ldots,$$
$$x_5 = 1.25992105\ldots.$$

At this point, we have the repetition of our four decimals, therefore, we can stop the process; it is $2^{1/3} = 1.2599$.

## 10.5 Numerical Integration: The Rectangular Rule

Many definite integrals cannot be done in closed form. For this reason, we need some useful techniques for approximating them. In the case of the 'rectangular rule', we consider the interpretation of a definite integral as the area under the curve $y = f(x)$ between $x = a$ and $x = b$:

$$\int_a^b f(x)dx = \sum_{i=1}^{n} f(\alpha_i)(x_i - x_{i-1}), \tag{10.7}$$

with $x_{i-1} \leq a_i \leq x_i$ and $a = x_0 < x_1 < x_2 < \ldots < x_n = b$. We obtain a good approximation of the integral dividing the interval $[a, b]$ into $n$ sub-intervals of length $h = (b - a)/n$ and replacing the function $f(\alpha_i)$ in every sub-interval by a straight line, which connects the values at each head (or end or at the centre point) of the sub-interval (Fig. 10.4).

We obtain:

$$\int_a^b f(x)dx \approx h(y_0 + y_1 + \ldots + y_{n-1}), \tag{10.8}$$

with $y_0 = f(x_0), y_1 = f(x_1), \ldots\ldots, y_{n-1} = f(x_{n-1})$. We note that, increasing $n$, the error rapidly decreases.

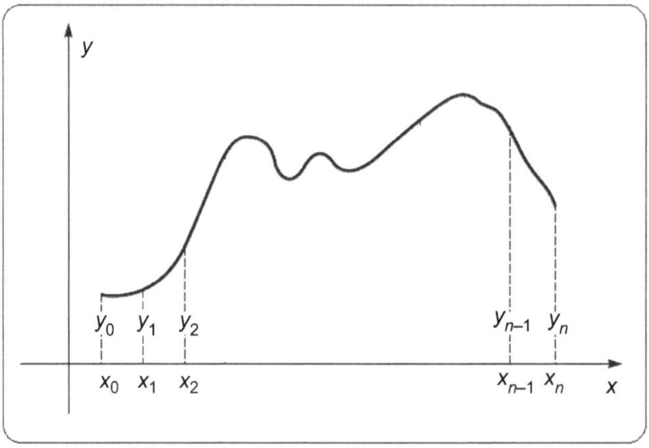


**Figure 10.4** The rectangular rule.

## 10.6 Numerical Integration: The Trapezoidal Rule

In this case, the small area of a sub-interval is slightly differently evaluated. We consider a trapezoid as shown in Fig. 10.5; its area is given by:

$$A = \frac{1}{2} h(y_1 + y_2). \tag{10.9}$$

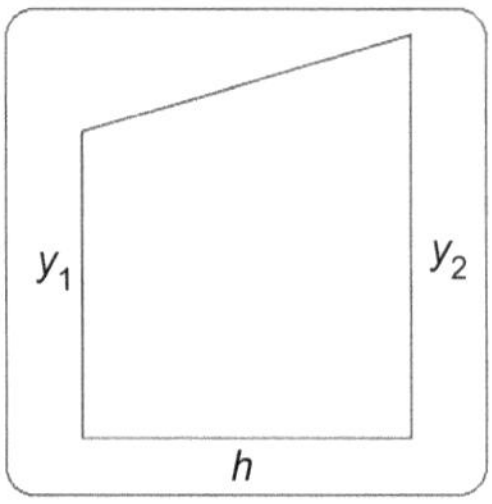


**Figure 10.5** The trapezoidal rule.

Applying Eq. 10.9, we obtain the approximation:

$$\int_a^b f(x)dx \approx \frac{(b-a)}{n}\left(\frac{1}{2}y_0 + y_1 + \ldots + y_{n-1} + \frac{1}{2}y_n\right). \tag{10.10}$$

About an upper and lower limit on the error of this method, starting with the calculation of the error for a single sub-interval of length $h = (b - a)/n$, we can write $x_i + h = z$ and $\varepsilon_i(z)$ for the error, obtaining:

$$\int_{x_i}^{z} f(x)dx = \frac{1}{2}h(y_i + y_z) + \varepsilon_i(z), \quad (10.11)$$

with $y_i = f(x_i)$ and $y_z = f(z)$. Through calculation, which involves a twice differentiation and anti-differentiation, if $m$ and $M$ are the minimum and the maximum of $f''(z)$ in the interval $[a, b]$, we get:

$$\frac{(b-a)^3}{12n^2}m \leq -\varepsilon \leq \frac{(b-a)^3}{12n^2}M. \quad (10.12)$$

Therefore, the error decreases rapidly as $n$ increases.

## 10.7 Numerical Integration: The Simpson Rule

In this case, we get a more accurate and useful formula of approximation for a definite integral. The equivalent of Eq. 10.10 becomes:

$$\int_{a}^{b} f(x)dx \approx \frac{h}{3}(y_0 + 4y_1 + 2y_2 + 4y_3 + 2y_4 + \ldots + 2y_{n-2} + 4y_{n-1} + y_n), \quad (10.13)$$

with $n$ even and $h = (b - a)/n$. It is possible to show that the error is inversely proportional to $n^4$. There are other methods, but more complicated, therefore, we refer to books of numerical analysis.

Chapter 11

# Quantum Basics for Nanotechnology

## 11.1 Black Body Radiation and Planck Hypothesis

A cold solid body does not produce any emissions, but with increasing temperature, it begins to become brighter and changes colour. A black body is a theoretical object, which absorbs 100% of the radiation that arrives on it. Therefore, it does not reflect any radiation and it appears perfectly black. In practise, any material absorbs all the incident radiation. A black body heated to a sufficiently high temperature emits radiation; the emitted energy depends only on the temperature of the body and not by its form or by the material from which it is made. The energy emitted by a black body heated to a temperature $T$ is called *black body radiation*.

An example of a black body is the *furnace*. The energy comes from a small hole; it is absorbed by the walls of furnace that heat up and emit radiation (Fig. 11.1).

The study of the properties of black body radiation played a key role in the development of quantum physics. It has led to the introduction of the Planck's constant $h$ and to the development, by Einstein, of the complex issue of what will be later referred as *wave–particle duality*.

*Mathematics and Physics for Nanotechnology: Technical Tools and Modelling*
Paolo Di Sia

ISBN 978-981-4800-02-0 (Hardcover), 978-0-429-02775-8 (eBook)
www.panstanford.com

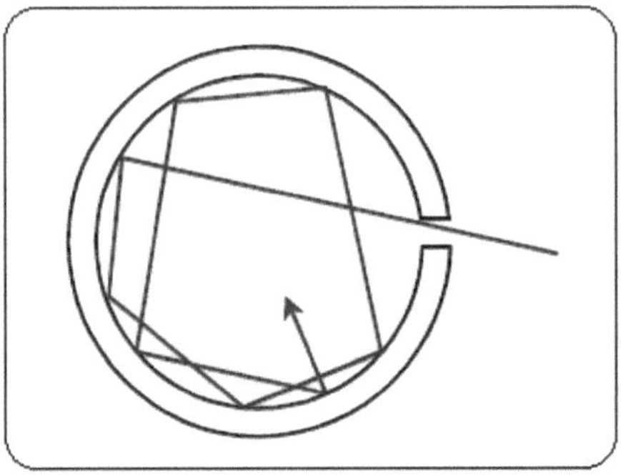

**Figure 11.1** Example of black body.

The history of the black body begins around 1860 when Kirchhoff defines what is meant by black body and formulates some general laws on its properties. Kirchhoff made the hypothesis according to which a body is able to absorb radiation that emits, thus explaining black lines observed by Fraunhofer and allowing to draw conclusions on the composition of the solar atmosphere. Kirchhoff has shown that 'at a given temperature and for a given wavelength, the ratio of the emissive power and that of absorption is the same for all bodies':

$$\frac{e(\lambda,T)}{a(\lambda,T)}=f(\lambda,T), \tag{11.1}$$

where $e$ is the emissive power, $a$ is the absorption power and $f$ is the universal function, independent by the nature of the considered body.

In 1879, based on experimental data of other authors, Stefan argued that the total energy emitted by a black body is proportional to $\sigma T^4$, with $\sigma$ constant.

In 1893, Wien combined the electromagnetism and thermodynamics, showing that the energy density of radiation in an isotherm cavity is:

$$u(\nu,T)=\nu^3 f\left(\frac{\nu}{T}\right). \tag{11.2}$$

In 1896, Wien published an article based on certain assumptions, coming to the conclusion that:

$$u(\nu,T)=a\nu^3 \exp\left(-\frac{b\nu}{T}\right). \tag{11.3}$$

This law had a satisfactory agreement with experimental data.

Planck began to work on a cavity model where the walls were composed of resonators, i.e., by electrically charged harmonic oscillators capable of exchanging energy with the electromagnetic radiation. Relying only on the laws of electromagnetism, Planck derived the formula:

$$u(\nu,T)=\frac{8\pi\nu^2}{c^3}\bar{U}_\nu, \tag{11.4}$$

which gives the energy density of the radiation in the cavity as a function of the average vibrational energy of the resonator. Considering Eqs. 11.2 and 11.4, and considerations about entropy and thermodynamics, Planck obtained the relation:

$$u(\nu,T)=\frac{A\nu^3}{e^{B\nu/T}-1}. \tag{11.5}$$

The Planck's function relative to the intensity of energy at various frequencies for the radiation emitted by a body at temperature $T$ is given by (Fig. 11.2):

$$f(\nu,T)=\frac{2h\nu^3}{c^2}\frac{1}{\exp\left(\frac{h\nu}{k_B T}\right)-1}. \tag{11.6}$$

The idea of Planck was as follows: the walls of a cavity, such as any emitting surfaces, contain particles, that absorbing energy from outside increase their temperature, therefore their kinetic energy and begin to oscillate. While oscillating they emit radiation, but this radiation, in contrast to the classical principles, cannot take any values. Energy must be emitted in *defined quantities* or *packets*. At high frequencies (short wavelengths), the radiation must be emitted in bigger packages. If the particles do not have enough energy, we will not see the emission of high-frequency radiation. On the other hand, if the temperature increases, the particles will have enough energy to emit packets at gradually higher frequencies.

In 1905, Einstein confirmed the idea of Planck explaining the photoelectric effect and showing that the radiation is not only emitted, but also absorbed in the form of packets or *photons*.

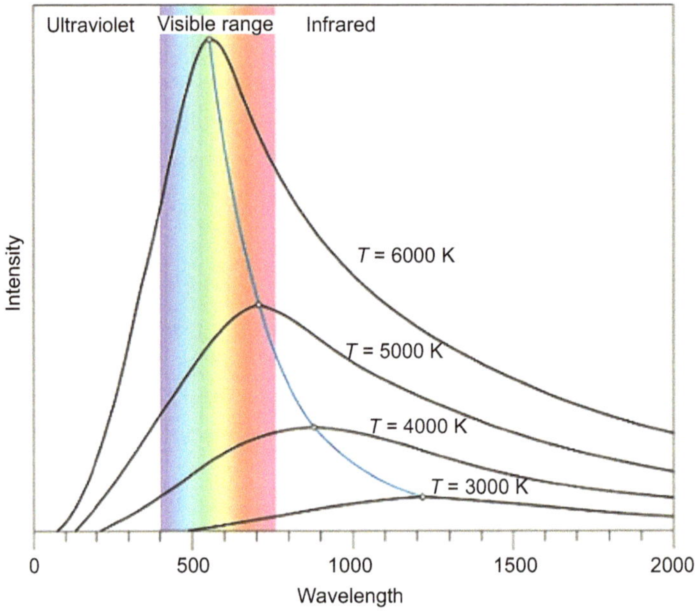


**Figure 11.2** Intensity of energy at various frequencies for radiation emitted by a body.

## 11.2 Einstein Work on Stimulated Emission

By the end of the nineteenth century, it was discovered that a beam of monochromatic light incident on a metal plate caused the emission of electrons (photoelectrons). Let us consider a metal as an electrically neutral object consisting of 'heavy' positive charges and negative charges (electrons), free to move inside the metal, but subjected to a potential that prevents its output under normal conditions. If the metal is heated or subjected to electromagnetic radiation, electrons can get a sufficiently high kinetic energy to allow to overcome the metal attraction potential and then to 'jump' out.

This phenomenon is called the *photoelectric effect* (Fig. 11.3).

The classical radiation theory can explain it, but not its main features. In 1905, Einstein proposed an explanation of the photoelectric effect assuming that electromagnetic radiation was emitted in packets, or *energy quanta*, called *photons*. A radiation of a frequency $\nu$ consists of photons of energy $h\nu$ and its intensity depends on the number of photons. The photon is completely absorbed by an electron, which increases its energy of $h\nu$.

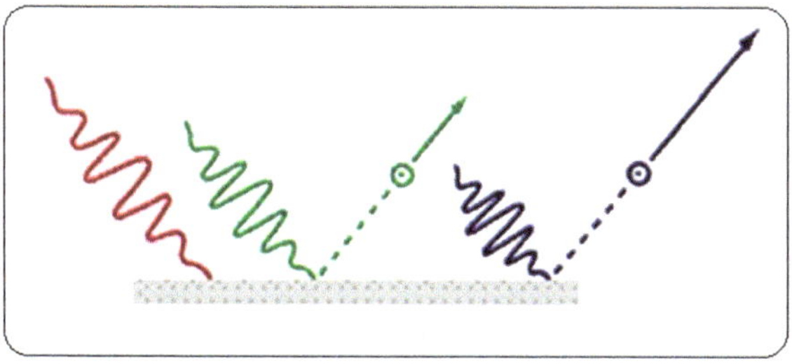

**Figure 11.3** Photoelectric effect.

When an electron that has absorbed a photon of frequency $\nu$ is emitted from the cathode with a kinetic energy $K$ given by:

$$K = h\nu - W, \tag{11.7}$$

with $W$ the required work to overcome the attraction of atoms and possible loss of kinetic energy due to impacts that must be provided for removing the electron from the metal; the maximum kinetic energy $K_{max}$ with which an electron can be emitted is given by:

$$K_{max} = h\nu - W_0 \geq 0, \tag{11.8}$$

where $W_0$, said *extraction work*, is a characteristic of the metal. Therefore, there exists a threshold frequency $h\nu_0 = W_0$, such that for $\nu < \nu_0$, Eq. 11.8 has no solution. The emission is immediate and does not depend on the intensity $I$ of radiation. Growing $I$, it increases the number of photons impacting on the cathode, thus increases the probability for the electrons to absorb a photon and, therefore, the number of emitted electrons.

The photoelectric effect provides a test, which is independent of the black body radiation, that electromagnetic radiation consists of quanta of energy $h\nu$.

## 11.3 De Broglie Waves

In 1924, Louis de Broglie proposed the existence of waves associated with particles. He postulated that each particle with impulse $p$ and energy $E$ is associated with a wavelength:

$$\lambda = h/p, \tag{11.9}$$

and a frequency:

$$\nu = E/h. \tag{11.10}$$

$\lambda$ is said the *de Broglie wavelength* of the particle. However, Eqs. 11.9 and 11.10 lead to an apparent contradiction with one of the postulates of special relativity of Einstein on the speed of light as maximum attainable value (Fig. 11.4).

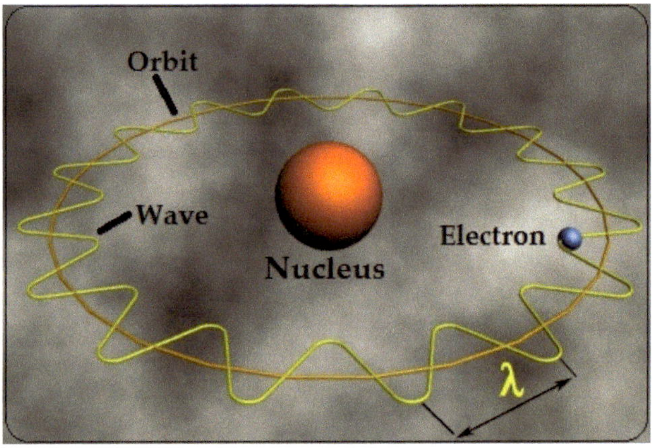


**Figure 11.4** De Broglie wavelength of a particle.

The apparent paradox disappears if we do not evaluate the *phase velocity* $v_{ph} = \lambda\nu = \omega/k$ (with $\omega = 2\pi\nu$), but the *group velocity* $v_g = d\omega/dk$ (with $k = 2\pi/\lambda$), which is the speed with which the energy propagates in a wave packet. In the non-relativistic limit, the group velocity is approximately the particle velocity.

The experimental confirmation of the wave nature of matter has been obtained in 1927 by Davisson and Germer.

## 11.4 The Compton Effect

For explaining the photoelectric effect, Einstein attributed to electromagnetic radiation a corpuscular nature, which had a clear experimental confirmation in 1923 with the Compton experiments. The idea has been to send a beam of X-rays on a crystal and to measure the intensity of the scattered radiation as a function of its wavelength.

Compton observed that, although the incident radiation is essentially monochromatic of wavelength $\lambda$, in the scattered radiation it was present at a new wavelength $\lambda'$, differed by $\lambda$ for a quantity that is a function of the diffusion angle. This result contradicts the predictions of classical physics. Classically it is expected that the free electrons of the crystal, under the action of the oscillating electric field of the incident radiation, oscillate at the same frequency and then emit scattered radiation in all directions, but of the same frequency of the incident wave. Compton explained the experimental result assuming that the incident radiation was constituted by a beam of photons, treated as particles, of energy $E = \lambda v$ and impulse $p = E/c = h/\lambda$. The elastic collision of photons with electrons must satisfy the conservation laws of energy and impulse; from that Compton obtained a relation that has good agreement with the experimental data:

$$\Delta\lambda = \lambda' - \lambda = \lambda_C(1 - \cos\theta), \tag{11.11}$$

where $\boldsymbol{\lambda}_C$ is said *Compton wavelength* ($\boldsymbol{\lambda}_C = 2.43 \times 10^{-12}$m) (Fig. 11.5). The shift $\Delta\lambda$ of the wavelength depends on the observation angle $\theta$, not on the initial length $\lambda$ of the incident beam.

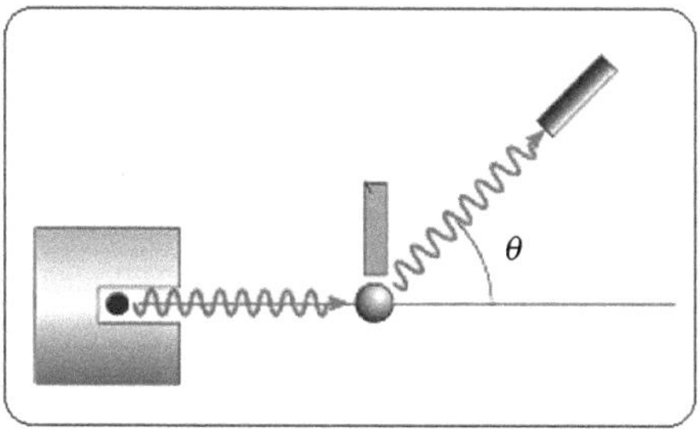


**Figure 11.5** Compton effect.

This experiment showed that the radiation is composed of photons that behave as particles, but at the same time the existence of an interference phenomenon requires a wave description of the radiation.

## 11.5 The Criterion Governing Classical and Quantum Properties of a Particle

The dimensionality of Planck's constant $h$ is [J s]. We can obtain the same dimensionality considering the products of these quantities:

(a) Time and energy;

(b) Coordinate and momentum;

(c) Angle and angular momentum.

The quantity with such a dimensionality is called *action* and normally indicated by $S$. So, Planck's constant can be said the *quantum of action*.

It follows that:

(a) If the action value of an object is *comparable* to the value of $h$, then its behaviour is governed by the quantum laws;

(b) If the action value of an object is *much larger* than $h$, then its behaviour is described with high precision by classical physics.

***Example* 11.1:** Calculate the action of a photon.

***Solution*:** The energy of a photon is equal to $E_{\text{ph}} = h\nu = \hbar\omega\, [\hbar = h/2\pi]$ and the characteristic time is equal to the period $T_{\text{ph}} = 2\pi/\omega$. The value of action for the photon is therefore $S_{\text{ph}} = E_{\text{ph}}T_{\text{ph}} = h$.

## 11.6 The Uncertainty Relations

In classical physics, the dynamic parameters of a particle (as coordinate, momentum, torque, energy, etc.) allow a complete description of its state. The real behaviour of microscopic particles shows that there is a limit 'in principle' to the precision with which these parameters can be measured. The uncertainty relations, suggested by Werner Heisenberg in 1927, determine the application limits of the behaviour of microscopic particles to the classical description.

The two most used expressions are:

(a) $\Delta x \Delta p_x \geq \hbar/2$ and similar relations for $y$ and $z$ coordinates in 3-D;

(b) $\Delta E \Delta t \geq \hbar/2$.

The dual wave–particle nature of microscopic objects is manifested by the impossibility of simultaneous exact measurement of the quantities involved in the uncertainty relations. The uncertainty principle, confirmed to date by available experimental data, allows us to establish the limits of application of classical theory; moreover thanks to it, we can obtain numerical estimates of parameters of studied systems without using complicated mathematical analysis.

## 11.7 The Reality at the Nanoscale and the Concept of Wave Function

The nanoworld is in the middle between classical physics and quantum physics, therefore, the use of this latter is often mandatory. The classical quantities for describing the motion of a particle are normally the position vector $\boldsymbol{r}$, the momentum $\boldsymbol{p}$, the energy $E$, etc. The state of a particle at a given time is defined by a number of physical quantities, described by a set of variables. In the case of a quantum-mechanical description, we need a smaller number of variables than that are necessary for a classical description.

Let us consider for example in classical physics the six variables $\boldsymbol{r}(t)$ and $\boldsymbol{p}(t)$ for the description of a particle's state. In the quantum case, we choose the coordinates $\boldsymbol{r}(t)$ as the total set of variables (such a description of the particle's state is called the *coordinate representation*). According to the uncertainty principle, the simultaneous measurement of coordinate and momentum of a particle is not possible; from the physical viewpoint, this is equivalent to the absence of a particle's trajectory, like in the classical way. If we talk about the probability of finding a particle in a given point at a given time, we must consider an infinitesimal volume $dV = dx\, dy\, dz$ including this point.

At the quantum level, we talk only about the probability of finding a certain value of a physical quantity, i.e., a *probabilistic description*. Some values of physical quantities are forbidden by quantum-mechanical laws; the allowed set is called the *spectrum*. It can be *discrete*, *continuous*, or *mixed*.

The *wave function* is a complex function $\Psi(\boldsymbol{r}, t)$, depending on time and coordinates, that is connected to the above indicated probability $P$ in this way:

$$dP = \Psi\Psi^* dV = |\Psi|^2 dV \,. \tag{11.12}$$

The physical meaning of the wave function resides in the square of its modulus $|\Psi|^2$, defined as the product of the wave function $\Psi$ and its complex conjugate $\Psi^*$.

The physical quantity:

$$\rho(\boldsymbol{r},t) = \frac{dP}{dV} = |\Psi|^2 \tag{11.13}$$

is the probability of finding the particle per volume unit; it is said *probability density* and is experimentally observable.

The finite probability of finding a particle within an arbitrary volume $V$ is given by:

$$P = \int_V \rho(\boldsymbol{r},t)dV = \int_V |\Psi|^2 dV \,. \tag{11.14}$$

Considering that the probability of finding a particle somewhere must be equal to 1, its wave function satisfies the *normalisation condition*:

$$\int_V |\Psi|^2 dV = 1 \,, \tag{11.15}$$

with the integral taken over the region where the wave function is different from zero. A wave function that satisfies the normalisation condition (Eq. 11.15) is said *normalised.*

One of the main postulates of quantum mechanics is the *superposition principle* for wave functions $\Psi$:

'If a particle can have two different quantum states described by the functions $\Psi_1$ and $\Psi_2$, then the particle can be in the state described by the function:

$$\Psi = c_1\Psi_1 + c_1\Psi_2, \tag{11.16}$$

with $c_1$ and $c_2$ two arbitrary complex numbers.'

In the general case of superposition of $N$ quantum states in which the particle can be, we write:

$$\Psi = \sum_{i=1}^{N} c_i \Psi_i . \tag{11.17}$$

The quantity $|c_i|^2$ defines the probability that during a measurement, we find a particle in a quantum state described by the wave function $\Psi_i$. If all wave functions in Eq. 11.17 are normalised, then the condition for the coefficient $c_i$ is:

$$\sum_{i=1}^{N} |c_i|^2 = 1 . \tag{11.18}$$

It holds the condition of *orthogonality* of wave functions:

'If a particle is in the state described by the function $\Psi_n$, the probability of finding it in the state described by another function $\Psi_m$ (with $m \neq n$) is zero.'

This condition, together with normalisation condition, has this form:

$$\int_V \Psi_m^*(\boldsymbol{r},t)\Psi_n(\boldsymbol{r},t)dV = \delta_{mn} , \tag{11.19}$$

with $\delta_{\mathrm{mn}}$ the Kronecker delta. The wave functions satisfying Eq. 11.19 are called *orthonormal* wave functions.

For a free particle, we can write the wave function in the form of de Broglie wave:

$$\Psi(\boldsymbol{r},t) = Ce^{-i(Et-\boldsymbol{p}\cdot\boldsymbol{r})/\hbar} . \tag{11.20}$$

If the wave function of a particle is a superposition of wave functions of free particles, then such a particle cannot be considered as free particle.

# Chapter 12

# Schrödinger Equation and Nanotechnology

## 12.1 The Schrödinger Equation

The Schrödinger equation is the main equation of non-relativistic quantum mechanics; with it we can find the wave function of a particle (or a system of particles in a stationary state) and also its evolution in time. In quantum mechanics, each physical quantity corresponds to its own operator; so this equation is written with operators.

An operator defines some mathematical manipulation for transforming a function into another function. These transformations are as follows:

(a) The multiplication of a function by a number or by another function on which the operator acts;

(b) The differentiation of the initial function;

(c) A shift of the function in time and/or in coordinate space, etc.

In the *coordinate representation*, the action of the coordinate operator on the wave function is equivalent to its multiplication by the corresponding coordinate.

*Mathematics and Physics for Nanotechnology: Technical Tools and Modelling*
Paolo Di Sia

ISBN 978-981-4800-02-0 (Hardcover), 978-0-429-02775-8 (eBook)
www.panstanford.com

*Examples*:

(a) The *momentum operator* in the coordinate representation is the differential operator given by:

$$\hat{p}=-i\hbar\nabla=-i\hbar\left(\boldsymbol{e}_x\frac{\partial}{\partial x}+\boldsymbol{e}_y\frac{\partial}{\partial y}+\boldsymbol{e}_z\frac{\partial}{\partial z}\right), \tag{12.1}$$

with $\boldsymbol{e}_x$, $\boldsymbol{e}_y$, $\boldsymbol{e}_z$ versors of axes of a coordinate system in 3-D.

(b) The operator of *kinetic energy* is based on the operator of the square of momentum:

$$\hat{p}^2=\hat{p}_x^2+\hat{p}_y^2+\hat{p}_z^2=-\hbar^2\nabla^2=-\hbar^2\left(\frac{\partial^2}{\partial x^2}+\frac{\partial^2}{\partial y^2}+\frac{\partial^2}{\partial z^2}\right). \tag{12.2}$$

(c) The operator of *angular momentum* is introduced considering the classical definition of the angular momentum, $\boldsymbol{L}=\boldsymbol{r}\times\boldsymbol{p}$ and substituting the expressions of operators for the coordinate and momentum. The expression, for a single particle, is given by:

$$\hat{L}=\hat{r}\times\hat{p}=-i\hbar\boldsymbol{r}\times\nabla. \tag{12.3}$$

The components are:

$$L_x=yp_z-zp_y=-i\hbar\left(y\frac{\partial}{\partial z}-z\frac{\partial}{\partial y}\right); \tag{12.4a}$$

$$L_y=zp_x-xp_z=-i\hbar\left(z\frac{\partial}{\partial x}-x\frac{\partial}{\partial z}\right); \tag{12.4b}$$

$$L_z=xp_y-yp_x=-i\hbar\left(x\frac{\partial}{\partial y}-y\frac{\partial}{\partial x}\right). \tag{12.4c}$$

(d) The operator of *total energy* is said *Hamiltonian* and, considering the expression for the energy $E$, is given by:

$$\hat{H}=\frac{\hat{p}^2}{2m}+\hat{U}(x,y,z)=-\frac{\hbar^2}{2m}\nabla^2+U(x,y,z). \tag{12.5}$$

For writing an operator equation, we introduce the notion of *eigenfunction* $\Psi$ of the operator $\hat{F}$ and *eigenvalues* $F$ of it. If the operator $\hat{F}$ corresponds to a physical quantity and its eigenfunction describes a possible state of the quantum system, taking values $F_n$, then the operator equation has the following form:

$$\hat{F}\Psi_n = F_n \Psi_n . \tag{12.6}$$

The operator $\hat{F}$ can have several (or infinitely many) eigenfunctions; the index $n$ corresponds to some states of the system. Some of the eigenvalues $F_n$ may correspond to several eigenfunctions $\Psi_{nm}$, which differ by the index $m$. Such eigenvalues $F_{nm}$ are said *degenerate*. The number of eigenfunctions with different indices $m$ corresponding to the given eigenvalue $F_a$ is said the *order of degeneracy of this eigenvalue*.

The operator corresponding to physical quantities must satisfy two conditions:

(a) It must be *linear* (*superposition* principle of quantum states):

$$\hat{F}(c_1\Psi_1 + c_2\Psi_2) = c_1\hat{F}\Psi_1 + c_2\hat{F}\Psi_2 . \tag{12.7}$$

(b) It must be *self-adjoint*, i.e., satisfying the integral condition:

$$\int_V \Psi_1^*(\hat{F}\Psi_2)dV = \int_V \Psi_2(\hat{F}\Psi_1)^* dV . \tag{12.8}$$

An operator with these characteristics has real eigenvalues corresponding to the physical quantity described by it.

(a) *Addition* and *product* of operators:

(i) Action of the *sum* of two operators on a wave function:

$$(\hat{F}_1 + \hat{F}_2)\Psi = \hat{F}_1\Psi + \hat{F}_2\Psi . \tag{12.9}$$

(ii) Action of the *product* of two operators on a wave function:

$$(\hat{F}_1\hat{F}_2)\Psi = \hat{F}_1(\hat{F}_2\Psi) = \hat{F}_1\Psi_2 = \Psi_{12} ; \tag{12.10}$$

$$(\hat{F}_2\hat{F}_1)\Psi = \hat{F}_2(\hat{F}_1\Psi) = \hat{F}_2\Psi_1 = \Psi_{21} . \tag{12.11}$$

If $\Psi_{12} = \Psi_{21}$, then the operators are said *commutative* ($\hat{F}_1\hat{F}_2 = \hat{F}_2\hat{F}_1$);

if $\Psi_{12} \neq \Psi_{21}$, then the operators are said *non-commutative* ($\hat{F}_1\hat{F}_2 \neq \hat{F}_2\hat{F}_1$).

***Example* 12.1:** Evaluate the commutativity of the operators $\hat{F}_1 = \hat{x} = x$ and $\hat{F}_2 = \hat{p}_x = -i\hbar\partial/\partial x$ .

***Solution***: We have:

$$(\hat{F}_1 \hat{F}_2)\Psi = -i\hbar\left(x\frac{\partial}{\partial x}\right)\Psi = -i\hbar x\frac{\partial \Psi}{\partial x} ;$$

$$(\hat{F}_2 \hat{F}_1)\Psi = -i\hbar\frac{\partial}{\partial x}(x\Psi) = -i\hbar\left(\Psi + x\frac{\partial \Psi}{\partial x}\right) .$$

Therefore, the two given operators do not commutate.

One of the most important postulates of quantum theory states:

If two different physical quantities $f_1$ and $f_2$ cannot be simultaneously measured, then the corresponding operators are non-commutative.

(b) *Average value* of a physical quantity in a given quantum state: It is one of the most important characteristic of a physical quantity, mathematically expressed as:

$$\langle F\rangle = \int_V \Psi^*(\boldsymbol{r},t)\hat{F}\Psi(\boldsymbol{r},t)dV , \tag{12.12}$$

for a state of the physical quantity $F$ (with the correspondent operator $\hat{F}$) described by $\Psi$ . The integration considers the entire volume of the motion of particle. In Cartesian coordinates, it is: $dV = dx\, dy\, dz$; furthermore, if $\hat{F} = F(\boldsymbol{r})$, then Eq. 12.12 becomes:

$$\langle F\rangle = \int_V F(\boldsymbol{r})|\Psi(\boldsymbol{r})|^2 dV . \tag{12.13}$$

***Example* 12.2:** Evaluate if the operator of complex conjugation $\hat{C}$ is linear.

***Solution***: We must verify if Eq. 12.7 holds. We consider $\hat{C}(c_1\Psi_1 + c_2\Psi_2)$; it is:

$$\hat{C}(c_1\Psi_1 + c_2\Psi_2) = c_1^*\Psi_1^* + c_2^*\Psi_2^* = c_1^*(\hat{C}\Psi_1) + c_2^*(\hat{C}\Psi_2) ;$$

this is different by: $c_1(\hat{C}\Psi_1) + c_2(\hat{C}\Psi_2)$, therefore, the considered operator is not linear.

## 12.2 The Time-Dependent Schrödinger Equation

This equation describes the behaviour of a non-relativistic particle governed by the laws of quantum mechanics:

$$\hat{H}\Psi(\boldsymbol{r},t)=i\hbar\frac{\partial}{\partial t}\Psi(\boldsymbol{r},t), \tag{12.14}$$

with $\hat{H}$ Hamiltonian of the system and $\Psi$ wave function describing it. We focus on some important relations:

(a) The normalisation condition of the wave function is *conserved* over time:

$$\int_V |\Psi|^2 dV = k\;; \tag{12.15}$$

(b) It holds the *continuity equation*:

$$\frac{d\rho}{dt}+\nabla\cdot\boldsymbol{J}=0, \tag{12.16}$$

where $\rho = |\Psi|^2$ is the *probability density* and:

$$\boldsymbol{j}=\frac{i\hbar}{2m}(\Psi\nabla\Psi^* - \Psi^*\nabla\Psi) \tag{12.17}$$

is the *probability current density.*

## 12.3 Stationary States

With stationary states, the physical quantities do not change with time. In this case, the dependence of the wave function on time is included in a phase factor:

$$\Psi(\boldsymbol{r},t)=\Psi(\boldsymbol{r})e^{-iEt/\hbar}\;. \tag{12.18}$$

The exponential factor, depending on time, changes only the phase of the wave function. So, the probability of finding a system in the corresponding state with energy $E$ does not depend on time:

$$|\Psi(\boldsymbol{r},t)|^2 = |\Psi(\boldsymbol{r})|^2. \tag{12.19}$$

The *stationary Schrödinger equation* is given by:

$$\hat{H}\Psi(\boldsymbol{r})=E\Psi(\boldsymbol{r}). \tag{12.20}$$

The eigenvalues of the operator $\hat{H}$ are some values $E_n$. The *ground state* is the state with the minimal energy $E_1 = E_{\min}$, similarly to the classical state of rest, for which $E_{\min} = 0$. Considering Eq. 12.5, Eq. 12.20 can be written as:

$$\left[-\frac{\hbar^2}{2m}\nabla^2 + U(\boldsymbol{r})\right]\Psi(\boldsymbol{r}) = E\Psi(\boldsymbol{r}). \qquad (12.21)$$

The solutions $\Psi(\boldsymbol{r})$ and the corresponding values of energy $E_i$ are defined in relation to the particular form of the potential energy $U(\boldsymbol{r})$.

## 12.4 Schrödinger Equation for Quantum Wires

If we consider a typical nanostructure, the confinement of electrons (or holes) leads to a reduction in dimensionality and therefore to a considerable change in its characteristics. The electronic environment is dimensionally reduced from a 2-D quantum well to a 1-D quantum wire and to a 0-D quantum dot, where dimensionality is related (in this context) to the number of freedom degrees of the electron moment (Fig. 12.1).

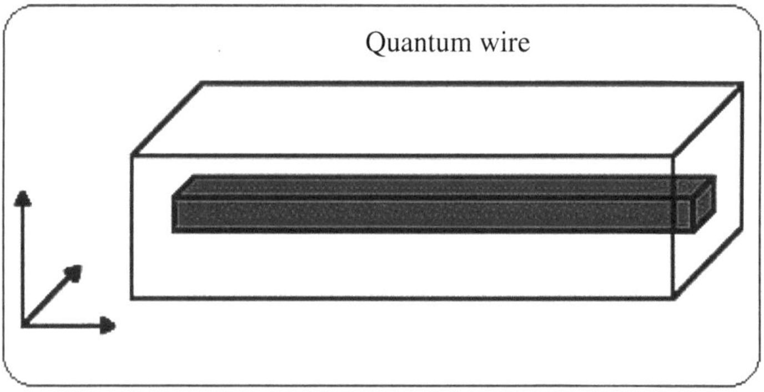


**Figure 12.1** Schematisation of a quantum wire.

Denoting with $D_\text{f}$ the number of degrees of freedom and with $D_\text{c}$ the number of confinement directions, for a solid state system it holds (Table 12.1):

$$D_\text{f} + D_\text{c} = 3. \qquad (12.22)$$

In the case of a quantum wire, the general 3-D Schrödinger equation for constant effective mass $m^*$ is given by:

$$-\frac{\hbar^2}{2m^*}\nabla^2\Psi(x,y,z)+V(x,y,z)\Psi(x,y,z)=E\,\Psi(x,y,z). \quad (12.23)$$

We can decouple the motion along the length of the wire; if $x$ is the axis of the wire, it is possible to write the potential as a sum of a 2-D confinement potential with the potential along the wire:

$$V(x,y,z)=V(x)+V(y,z). \quad (12.24)$$

**Table 12.1** Degrees of freedom and confinement directions in nanometric systems

| System | $D_c$ | $D_f$ |
|---|---|---|
| Bulk | 0 | 3 |
| Quantum well | 1 | 2 |
| Quantum wire | 2 | 1 |
| Quantum dot | 3 | 0 |

Writing the eigenfunction as the product of two components:

$$\Psi(x,y,z)=\Psi(x)\Psi(y,z), \quad (12.25)$$

and the energy as a sum of terms associated with the two motion components:

$$E=E_x+E_{y,z}, \quad (12.26)$$

we obtain two decoupled equations:

$$-\frac{\hbar^2}{2m^*}\Psi(y,z)\frac{\partial^2\Psi(x)}{\partial x^2}+\Psi(y,z)V(x)\Psi(x)=\Psi(y,z)E_x\,\Psi(x); \quad (12.27)$$

$$-\frac{\hbar^2}{2m^*}\left[\Psi(x)\frac{\partial^2\Psi(y,z)}{\partial y^2}+\Psi(x)\frac{\partial^2\Psi(y,z)}{\partial z^2}\right]+\Psi(x)V(y,\ z)\Psi(y,\ z)$$

$$=\Psi(x)E_{y,\,z}\Psi(y,\ z). \quad (12.28)$$

Being null the component of the potential along the axis of wire, Eqs. 12.27 and 12.28 reduce to:

$$-\frac{\hbar^2}{2m^*}\frac{\partial^2\Psi(x)}{\partial x^2}=E_x\,\Psi(x); \quad (12.29)$$

$$-\frac{\hbar^2}{2m^*}\left[\frac{\partial^2\Psi(y,z)}{\partial y^2}+\frac{\partial^2\Psi(y,z)}{\partial z^2}\right]+V(y,z)\Psi(y,z)=E_{y,z}\Psi(y,z)\,. \tag{12.30}$$

Equation 12.29 is satisfied by a plane wave of the form:

$$\Psi(x)=\exp(ik_x x)\,, \tag{12.31}$$

which gives:

$$E_x=\frac{\hbar^2 k_x^2}{2m^*}\,. \tag{12.32}$$

Equation 12.30 is the Schrödinger equation for the 2-D confinement potential characterising a quantum wire.

The *infinitely deep rectangular wire* can be regarded as a 2-D similar situation of the 1-D confinement potential of the standard infinitely deep quantum well. Considering that $V_{out} = \infty$ (thus $\Psi = 0$), the Schrödinger equation is defined only within the wire for the motion in the two confined directions $y$ and $z$. Equation 12.30 becomes:

$$-\frac{\hbar^2}{2m^*}\left[\frac{\partial^2\Psi(y,z)}{\partial y^2}+\frac{\partial^2\Psi(y,z)}{\partial z^2}\right]=E_{y,z}\Psi(y,z)\,. \tag{12.33}$$

Because of motion decoupling, we can write the eigenfunction as:

$$\Psi(y,z)=\Psi(y)\Psi(z), \tag{12.34}$$

and associate the terms of individual kinetic energy with separated energy components. Two equations are as follows:

$$-\frac{\hbar^2}{2m^*}\frac{\partial^2\Psi(y)}{\partial y^2}=E_y\Psi(y)\,; \tag{12.35}$$

$$-\frac{\hbar^2}{2m^*}\frac{\partial^2\Psi(z)}{\partial z^2}=E_z\Psi(z)\,. \tag{12.36}$$

Being $V_{out} = \infty$, the standard condition of contour continuity brings to $\Psi(y)=\Psi(z)=0$ at the wire border; Eqs. 12.35 and 12.36 are therefore 1-D equations, similar to that of the 1-D infinitely deep well. Considering the origin at a vertex of the rectangular section and said $L_y$ and $L_z$ the sizes of the wire, it holds:

$$\Psi(y) = \sqrt{\frac{2}{L_y}}\sin\left(\frac{\pi n_y y}{L_y}\right), \tag{12.37}$$

$$\Psi(z) = \sqrt{\frac{2}{L_z}}\sin\left(\frac{\pi n_z z}{L_z}\right), \tag{12.38}$$

so that:

$$E_y = \frac{\hbar^2\pi^2 n_y^2}{2m^*\ L_y^2}; \tag{12.39}$$

$$E_z = \frac{\hbar^2\pi^2 n_z^2}{2\ m^*\ L_z^2}. \tag{12.40}$$

The total energy, due to confinement, results:

$$E_{y,z} = E_y + E_z = \frac{\hbar^2\pi^2}{2\ m^*}\left(\frac{n_y^2}{L_y^2} + \frac{n_z^2}{L_z^2}\right), \tag{12.41}$$

with $n_y$ and $n_z$ the two principal quantum numbers.

In the configuration *with finite barriers*, it is not possible to separate the motion along $y$ from that along $z$ and writing the potential $V(y, z)$ as the sum of two independent parts. An approximate form of the potential in this case is specified in Table 12.2 for a rectangular wire.

**Table 12.2** Approximate form of potential $V(y, z)$ for a rectangular wire in confined directions $y$ and $z$

| | | |
|---|---|---|
| $2V$ | $V$ | $2V$ |
| $V$ | $V(y, z) = 0$ | $V$ |
| $2V$ | $V$ | $2V$ |

We can write $V(y, z) = V(y) + V(z)$, where $V(y)$ and $V(z)$ are the potentials of independent finite wells; it is a good approximation for long wires and for the lowest energy states.

Considering Eq. 12.30, writing the potential as indicated, the wave function as $\Psi(y,z)=\Psi(y)\Psi(z)$ and the energy $E_{y,z}$ as a sum of two components associated with the motions along $y$ and $z$, it is possible to split the equation into two parts. After calculation, we get:

$$-\frac{\hbar^2}{2m^*}\frac{\partial^2\Psi(y)}{\partial y^2}+V(y)\Psi(y)=E_y\Psi(y), \qquad (12.42)$$

$$-\frac{\hbar^2}{2m^*}\frac{\partial^2\Psi(z)}{\partial z^2}+V(z)\Psi(z)=E_z\Psi(z), \qquad (12.43)$$

i.e., the equations for the simple 1-D potentials. The eigenvalues of energy $E_{y,z}$ can be determined using the first order perturbation theory. This approach is applicable for a finite barrier for quantum wires and dots.

***Cylindrical wire***: In this case, the cylindrical symmetry is convenient and helps to use the polar coordinates $(r, \theta)$ and the wave function $\Psi$ $(y, z)$ can be written in terms of the new variables. Moreover, thanks to a further symmetry of circular type, the $\Psi$ function results independent from $\theta$. The Schrödinger equation for $\Psi(r)$ becomes:

$$-\frac{\hbar^2}{2m^*}\left[\frac{\partial^2}{\partial y^2}+\frac{\partial^2}{\partial z^2}\right]\Psi(r)+V(r)\Psi(r)=E_r\Psi(r), \qquad (12.44)$$

where the subscript $r$ in the energy term indicates that this eigenvalue is associated to the 'cross-sectional' confined motion, as opposed to the not confined motion along the wire axis. The circular symmetry allows to write the potential, that defines the wire, directly as $V(r)$. Deriving twice both members of the relation $r=\sqrt{y^2+z^2}$ with respect to $y$ and $z$ and inserting the obtained relation in Eq. 12.44, we have:

$$-\frac{\hbar^2}{2m^*}\left[\frac{1}{r}\frac{\partial}{\partial r}+\frac{\partial^2}{\partial r^2}\right]\Psi(r)+V(r)\Psi(r)=E_r\Psi(r), \qquad (12.45)$$

valid only for constant effective mass.

## 12.5 Schrödinger Equation for Quantum Dots

The *cubic quantum dot* can be considered as a generalisation of the quantum wire with rectangular section, with an additional confinement along the remaining $x$-axis, then removing the remaining degree of freedom in the moment of the particle. If an infinite potential separates the inside part from the outside part of the box, the 3-D Schrödinger equation inside the box is given by:

$$-\frac{\hbar^2}{2m^*}\left[\frac{\partial^2}{\partial x^2}+\frac{\partial^2}{\partial y^2}+\frac{\partial^2}{\partial z^2}\right]\Psi(x,y,z)=E_{x,y,z}\,\Psi(x,y,z)\,. \quad (12.46)$$

Writing the total energy as $E_{x,y,z} = E_x + E_y + E_z$ , it is possible to decouple Eq. 12.46 into three 1-D equations:

$$-\frac{\hbar^2}{2m^*}\frac{\partial^2}{\partial x^2}\Psi(x)=E_x\,\Psi(x)\,; \quad (12.47)$$

$$-\frac{\hbar^2}{2m^*}\frac{\partial^2}{\partial y^2}\Psi(x)=E_y\,\Psi(x)\,; \quad (12.48)$$

$$-\frac{\hbar^2}{2m^*}\frac{\partial^2}{\partial z^2}\Psi(x)=E_z\,\Psi(x)\,. \quad (12.49)$$

The energy confinement inside the box is:

$$E_{x,y,z}=\frac{\hbar^2\pi^2}{2m^*}\left(\frac{n_x^2}{L_x^2}+\frac{n_y^2}{L_y^2}+\frac{n_z^2}{L_z^2}\right). \quad (12.50)$$

Equation 12.50 requires three quantum numbers $n_x$, $n_y$, $n_z$ to define each state.

In the case of *spherical quantum dots*, for a given spherically symmetric potential (Fig. 12.2), we can use a wave function with spherical symmetry.

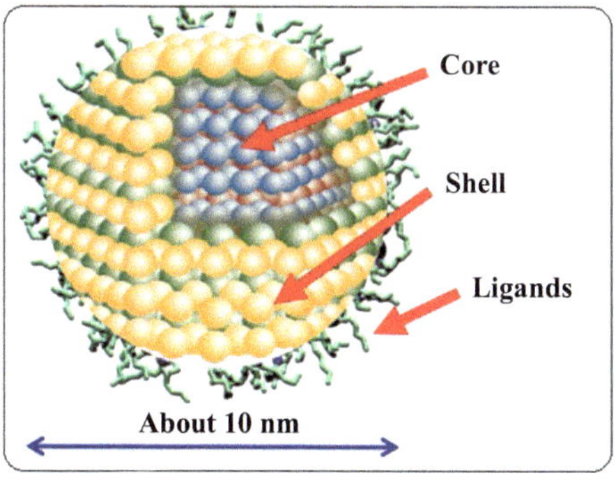


**Figure 12.2** Schematisation of a spherical quantum dot.

The Schrödinger equation for a constant effective mass is given by:

$$-\frac{\hbar^2}{2m^*}\left[\frac{\partial^2}{\partial x^2}+\frac{\partial^2}{\partial y^2}+\frac{\partial^2}{\partial z^2}\right]\Psi(r)+V(r)\Psi(r)=E_r\Psi(r), \qquad (12.51)$$

where $E_r(r)$ is the energy associated with the confinement along the radius.

Using spherical coordinates $(r,\theta,\varphi)$, we can write for each Cartesian axis an equation of the form:

$$\frac{\partial^2}{\partial x^2}\Psi(r)=\frac{1}{r}\frac{\partial}{\partial r}\Psi(r)-\frac{x^2}{r^3}\frac{\partial}{\partial r}\Psi(r)+\frac{x^2}{r^2}\frac{\partial^2}{\partial r^2}\Psi(r); \qquad (12.52)$$

$$\frac{\partial^2}{\partial y^2}\Psi(r)=\frac{1}{r}\frac{\partial}{\partial r}\Psi(r)-\frac{y^2}{r^3}\frac{\partial}{\partial r}\Psi(r)+\frac{y^2}{r^2}\frac{\partial^2}{\partial r^2}\Psi(r); \qquad (12.53)$$

$$\frac{\partial^2}{\partial z^2}\Psi(r)=\frac{1}{r}\frac{\partial}{\partial r}\Psi(r)-\frac{z^2}{r^3}\frac{\partial}{\partial r}\Psi(r)+\frac{z^2}{r^2}\frac{\partial^2}{\partial r^2}\Psi(r). \qquad (12.54)$$

Writing $\nabla^2\Psi(r)$ and replacing it in Eq. 12.51, we obtain:

$$-\frac{\hbar^2}{2m^*}\left[\frac{2}{r}\frac{\partial}{\partial r}+\frac{\partial^2}{\partial r^2}\right]\Psi(r)+V(r)\Psi(r)=E_r\Psi(r), \qquad (12.55)$$

i.e., the Schrödinger equation with spherical symmetry.

## 12.6 Schrödinger Equation and Drude–Lorentz-like Model

The transport processes of carriers in a nanostructure are well described by the Drude–Lorentz model, as improvement of the Drude model and its variants, like the Smith model and the effective medium theories (EMTs), which incorporate the Maxwell–Garnett model (MG) and the Bruggeman model (BR).

Another recently appeared generalisation is an interesting model (said DS model) based on the complete Fourier transform of the frequency-dependent complex conductivity $\sigma(\omega)$ of the system. It brings to analytical expressions of the most important quantities related to transport phenomenon, i.e., the velocity correlation function $<\boldsymbol{v}(t) \cdot \boldsymbol{v}(0)>_T$ at the temperature $T$, the mean square deviation of position $R^2(t) = \langle [\boldsymbol{R}(t) - \boldsymbol{R}(0)]^2 \rangle$ and the diffusion coefficient $D$, avoiding time-consuming numerical procedures and particular simulations. The model is able both to study new 'a priori' characteristics of nanomaterials and nano-based devices and at 'a posteriori' level, testing existing experimental data. It demonstrated good agreement with the theoretical results of past models, as well as in relation to the existing data fits. The model has been performed in the classical, quantum and relativistic case (see Chapters 13–15).

The parameter $\omega_0$ of the Drude–Lorentz model corresponds to the oscillator frequency and describes the binding of particles to atoms of the system. It can be obtained by experimental techniques, such as the well-known 'time-resolved THz spectroscopy' (TRTS), a non-contact electrical probe technique, able to measure the photoconductivity from sub-picoseconds to nanoseconds.

Another way of evaluation is obtainable from the binding energy of the charges in nanoparticles composing the nanostructure, from the excited energy for a potential well of nanosize $x$, through the AFM analysis of the nanostructure. Considering the approximately spherical form of many substructures composing a nanocrystals (Figs. 12.3 and 12.4), we can deal with a spherical potential well.

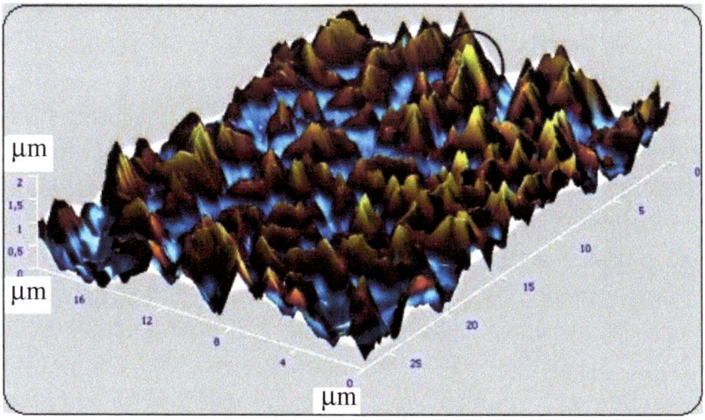


**Figure 12.3** AFM image of a microstructure of ZnO.

In the circle of Fig. 12.3, we note the typical pyramidal local form of the structure. Going down to the nanoscale, the tips of the pyramids become rounded (Fig. 12.4).

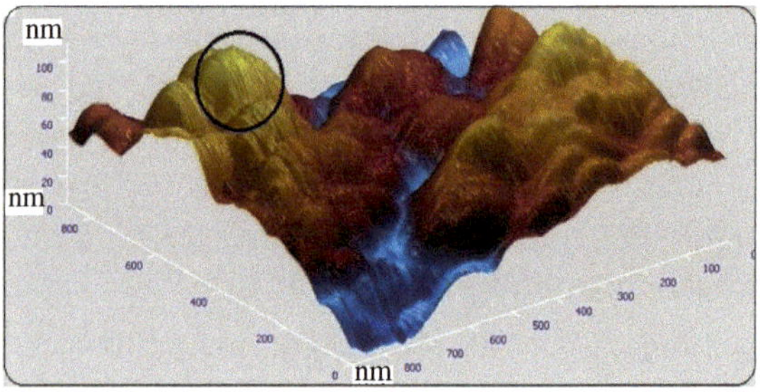


**Figure 12.4** AFM image of a nanostructure of $TiO_2$ with the typical spherical local form of the structure.

The energy levels $E_i$ of the system can be deduced from the results of quantum mechanics, using the Schrödinger equation. In the first step, it is possible to solve the eigenvalue equation, with the assumption of only one bound state and find the energy, which corresponds to a value of $\omega_0$. Analogous calculation can be performed for chosen potentials in relation to the form of nanoparticles, as nanotubes, nanorods, nanostrings, nanobelts, etc.

## 12.7 Schrödinger Equation and Computational Nanotechnology

The *computational nanotechnology* is a way for numerically studying the development of current and next generations of multifunctional electronics and computing, nanosensors, actuators, machines and nanodevices. The length and time scales of nanoscale systems make possible a direct study through computer simulations and theoretical modelling with high accuracy. The increasing power of computers, currently used at large scale and with high fidelity, allows accurate predictive nanoscale simulations. The computational nanotechnology emerges as a fundamental tool of engineering analysis for nanodevice design in the way that continuum finite element analysis has utilised for designing and analysing most engineering systems in previous years.

Many areas can be analysed with the computational nanotechnology, as:

(a) molecular-scale, ultralight, strong, functional materials;
(b) nanoscale electronics with possibilities for quantum computing;
(c) molecular-scale sensors and actuators;
(d) molecular machines and motors with synthetic materials.

In the recent past, researchers computed the thermodynamic properties of interacting bulk condensed-matter systems through analytical approximation methods for an infinite system, with validity related to the weakly interacting system limit. Then new types of approximation for studying interacting condensed-matter systems have been studied, in particular 'exact numerical computation' of the properties of finite-sample systems. The most general approaches describe atoms as a collection of quantum mechanical particles, nuclei and electrons, governed by the Schrödinger equation. Some methods, such as the 'ab-initio' method, solve complex quantum many-body Schrödinger equations with numerical algorithms, preferring the linear combination of atomic orbital approximations in describing the quantum mechanical electronic wave functions.

The 3-D simulations of nanoelements involve a self-consistent solution of a 3-D Poisson equation and a 3-D Schrödinger equation with open boundary conditions. Using the finite element method (FEM), the resolution of the 3-D Schrödinger equation can provide both computational efficiency and high accuracy.

With the 'mode space approach', based on an expansion of the active device Hamiltonian in the sub-band eigenfunction space, the calculation is reduced to the resolution of a 2-D Schrödinger equation, for obtaining the sub-band eigenenergy levels and eigenfunctions (modes). For increasing the efficiency of the simulator, only 2-D Schrödinger equations can be involved in a self-consistent loop, providing great computational accuracy.

With these methods, computationally efficient 3-D quantum simulations of various nanowire transistors based on the effective-mass approximation can be done. Particular approaches, such as the coupled or uncoupled mode space approaches, allow to decompose the 3-D device Hamiltonian, that greatly reduces the simulation time and keeps high computational accuracy.

# Chapter 13

# Mathematical Modelling for Nanotechnology

## 13.1 Introduction

One of the most important aspects at nanoscale concerns the charge transport, which can be influenced by particles dimensions and assume different characteristics with respect to those of bulk. In particular, if the mean free path of charges, due to scattering phenomenon, is larger than the particle dimensions, we have a mesoscopic system, in which the transport depends on dimensions and in principle it is possible to correct the transport bulk theories by considering this phenomenon. A similar situation occurs also in a thin film, in which the smallest nanostructure dimension can be less than the free displacement and therefore require variations to existing theoretical transport bulk models. Therefore, a rigorous knowledge of transport properties is to be acquired. From a theoretical point of view, various techniques can be used for the comprehension of transport phenomenon; particularly powerful are the analytical descriptions based on transport equations. Currently many used theories consider numerical approaches, but analytical results are of great mathematical interest and suitable to be implemented through experimental data from the literature. Therefore, new theoretical approaches are welcome, in particular

*Mathematics and Physics for Nanotechnology: Technical Tools and Modelling*
Paolo Di Sia

ISBN 978-981-4800-02-0 (Hardcover), 978-0-429-02775-8 (eBook)
www.panstanford.com

for having analytical expressions of the most important transport functions, i.e., the velocities correlation function, the mean square deviation of position and the diffusion coefficient.

At the nanoscale, we are in the middle between classical and quantum effects, macroscopic and microscopic properties of the matter. Quantum effects in the nanoworld have opened new ways in many old and new technological sectors. About the experimental aspects, currently one of the most important experimental technique for the study of the frequency-dependent complex-valued far-infrared photoconductivity $\sigma(\omega)$ is the time-resolved THz spectroscopy (TRTS), an ultrafast non-contact optical probe; obtained data are normally fitted using Drude–Lorentz, Drude–Smith or effective medium models. In the following, I illustrate an overview of fundamental models that, starting from the Drude model, analyse and explain the transport phenomenon of the matter at the nanoscale, coming to recent developments concerning variations of Drude–Lorentz-like models.

## 13.2 The Drude Model

The most meaningful characteristics of metals are their properties of electric and thermal conduction. Over the years, this fact brought to think in terms of models in which the electrons are relatively free and can move under the influence of electric fields. Historically, two important models of the elementary theory of metals are born:

(a) The *Drude model*, published in 1900 and based on the kinetic theory of an electron gas in a solid; it is assumed that all electrons have the same average kinetic energy $E_m$.

(b) A variation of the previous one, integrated with the foundations of quantum mechanics, called *Sommerfeld model*.

In the Drude model, the valence electrons of atoms are considered free inside the metal; all electrons move as an electronic gas. The ideal gas equations are applied, even if the electron gas is dense. If the electrons try to leave the metal, they are retained by the net positive charge left on the metal; they are then in a potential energy hole of depth $W$. The uniform distribution of positive charges associated with the metal ions, that make the metal neutral ‘on average', determine a large value of $W$. The work $L = W - E_m$ has an enough large value to retain the electrons in the solid. The kinetic theory leads to an average energy medium $k_B T/2$ for each degree of freedom ($k_B$ = Boltzmann's constant).

Free electrons have kinetic energy and no potential energy. It is assumed that all electrons have the same energy $E_m$ and it exists a mechanism of collisions between ions and electrons, allowing the thermal equilibrium for the electrons. This implies the possibility to apply the kinetic theory of gases to such electronic gas. Drude published the theory three years after the discovery of the electron by J. J. Thomson. The average energy $E_m$ can be correlated to an average quadratic speed $v_m$ through the relation:

$$E_m = (3/2)k_B T = m v_m^2/2, \tag{13.1}$$

with $m$ the mass of the free electron. *At environment temperature,* $v_m \cong 10^7$ cm/s represents the average thermal speed of the electrons. It was assumed also that the electrons have collisions as instant events, i.e., the time of the diffraction is considered very small with respect to every other involved time. Through such collisions, electrons acquire a thermal equilibrium corresponding to the temperature $T$ of the metal. If the electrons do not collide, they move in linear way following the Newton laws. The presence of a constant electric field determines an extra average velocity (the *drift velocity*) given by:

$$v_d = -(eE/m)t. \tag{13.2}$$

The *relaxation time* $\tau$ is defined as the average time between two collisions; it is possible to get a mean free path ($l_{mfp}$), defined by:

$$l_{mfp} = v_m \tau. \tag{13.3}$$

On the basis of these assumptions, it is possible to obtain many properties of an electron gas. The fact that $v_d << v_m$ is consistent with the idea that the current flow is only a small perturbation in the normal behaviour of electrons.

The current density is $\boldsymbol{J} = \sigma_{cond} \boldsymbol{E}$, with $\sigma_{cond}$ electric conductivity. This result, obtained by Drude, is an important goal of the classic theory on the conduction of metals, said *Drude theory*.

By simple electrical conductivity measurements, it is possible to obtain the value of $\tau$, estimating the density of conduction electrons. For a metal, it is $\sigma_{cond} \approx 10^8\ \Omega^{-1}\ \text{cm}^{-1}$ and assuming an electron density of the order of $10^{23}$ electrons/cm$^3$, we get $\tau \approx 10^{-14} - 10^{-15}$ s. To that it corresponds mean free path $l_{\text{mfp}} \approx v\,\tau \approx 1 - 10$ Å, that is, the order of the interatomic distances in a crystal lattice.

## 13.3 The Drude–Lorentz Model

The Lorentz model (1905) is a refining of the Drude model, in which the statistical aspects are specified. Electrons are considered as free charges, with charge $-e$; they are described by a Maxwellian velocity distribution. Considering an electron gas in a spatial region with a constant electric field, the drift velocity of electrons is constant; this corresponds to a current density $\boldsymbol{J}$ proportional to the applied field $\boldsymbol{J} = \sigma_0\,\boldsymbol{E}$, with $\sigma_0 = ne^2\,\tau/m$ ($n$ is the electron density).

Estimating the relaxation time $\tau$, Drude and Lorentz obtained values of conductivity in good agreement with the experiments. In presence of an electric field of the form $E(t)=E_0\,e^{-i\omega t}$, the complex conductibility results:

$$\sigma_\omega = \sigma_0/(1-i\omega\tau). \qquad (13.4)$$

Such model, said *Drude–Lorentz model*, received great success during years, but has also underlined big problems.

## 13.4 About the Most Utilised Drude–Lorentz-like Models

The Drude–Lorentz model is one of the most used models for describing data concerning the charge transport. Through this model, it is possible to get the most important parameters of the transport, i.e.,

(a) The *velocities correlation function*, by which it is possible to get the speed of charges.

(b) The *mean square deviation of position* as a function of time, which allows to get the space travelled by charges.

(c) The *diffusion coefficient*.

An extension of this model has been proposed by Smith and subsequently applied to fit the conductivity in a wide variety of systems. In Smith model, the complex conductivity is written through a parameter that takes into account the anisotropy of scattering after the first scattering event. The additional terms with respect to the Drude model correspond to a backscattering of the charge carriers.

Considerable variations of this model have been made; the most used are as follows:

(a) **The *Maxwell–Garnett model* (MG)**: In this model, the dielectric function is given by a Drude term with an additional 'vibrational' contribution at a finite frequency $\omega_0$, leading to a dielectric function of the form:

$$\varepsilon_{//}(\omega) = 1 - \frac{\omega_{\mathrm{p}}^2}{\omega(\omega + i/\tau)} + \frac{\omega_{\mathrm{s}}^2}{\omega_0^2 - \omega^2 - i\gamma\omega}, \qquad (13.5)$$

where the amplitude $\omega_{\mathrm{s}}$, the resonant frequency $\omega_0$ and the damping constant $\gamma$ are constants depending on the particular considered material. The MG model usually describes an isotropic matrix containing spherical inclusions isolated from each other, such as metal particles dispersed in a surrounding host matrix.

(b) **The *Bruggeman model* (BR):** It is used for the study of the effective conductivity of systems consisting of multi-component spherical inclusions with different conductivity. The Bruggeman formula for circular and spherical inclusions is the following:

$$\sum_i \delta_i \frac{\sigma_i - \sigma_{\mathrm{e}}}{\sigma_i + (n-1)\sigma_{\mathrm{e}}} = 0, \qquad (13.6)$$

where $n$ is the spatial dimension, the sum is performed on all constituents, $\delta_i$ (with $\sum_i \delta_i = 0$) and $\sigma_i$ are the fraction and the conductivity of each component, respectively, and $\sigma_{\mathrm{e}}$ the effective conductivity of the medium. For elliptical and ellipsoidal inclusions, the formula is:

$$\frac{1}{n}\delta\alpha + \frac{(1-\delta)(\sigma_{\mathrm{m}} - \sigma_{\mathrm{e}})}{\sigma_{\mathrm{m}} + (n-1)\sigma_{\mathrm{e}}} = 0. \qquad (13.7)$$

Equation 13.7 is a generalisation of Eq. 13.6 for a two-phase system with conductivity inclusions $\sigma_{\mathrm{e}}$ in a matrix of conductivity $\sigma_m$. The quantity $\alpha$, for randomly oriented inclusions, is given by:

$$\alpha = \sum_{j=1}^{n} \frac{\sigma - \sigma_e}{\sigma_e + L_j(\sigma - \sigma_e)}, \tag{13.8}$$

where $L_j$ (with $\sum_j L_j = 0$) denotes the appropriate doublets or triplets depolarisation factors, governed by the ratio between the axes of the ellipse or ellipsoid.

(c) ***Effective medium theories*** **(EMTs):** In this case, the electromagnetic interactions between pure materials and host matrices are approximately taken into account. The commonly used EMTs include the Maxwell-Garnett model and the Bruggeman model. Generally, in the THz regime, the dielectric function $\varepsilon_m(\omega)$ consists of contributions of the high-frequency dielectric constant, free electrons of conduction and lattice vibration. The dielectric function has the form:

$$\varepsilon_m(\omega) = \varepsilon_\infty - \frac{\omega_p{}^2}{\omega^2 + i\gamma\omega} + \sum_j \frac{\varepsilon_{st_j}\,\omega_{TO_j}{}^2}{\omega_{TO_j}{}^2 - \omega^2 - i\Gamma_j\omega}, \tag{13.9}$$

in which $\varepsilon_\infty$ is the high-frequency dielectric constant, the second term describes the contribution of free electrons or plasmons and the last term stands for the optical phonons.

When the response is mainly originated by the contribution of free electrons or plasmons, it is usually considered:

$$\varepsilon_m(\omega) = \varepsilon_\infty - \frac{\omega_p{}^2}{\omega^2 + i\gamma\omega}, \tag{13.10}$$

that well describes the dielectric properties of metals and semiconductors.

If instead the interaction of a radiation field with the fundamental lattice vibration plays a dominant role and results in absorption of electromagnetic wave, due to the creation or annihilation of lattice vibration, the dielectric function $\varepsilon_m(\omega)$ consists of the contributions of the lattice vibrations, expressed by the classical pseudo-harmonic phonon model in the first approximation:

$$\varepsilon_m(\omega)=\varepsilon_\infty+\frac{\varepsilon_{st}\,\omega_{\mathrm{TO}}^{\;2}}{\omega_{\mathrm{TO}}^{\;2}-\omega^2-i\gamma\omega}. \tag{13.11}$$

## 13.5 The Smith Model

The Smith model starts with the theory of response for the optical conductivity, considering a pulse of electric field applied to the system at the time $t = 0$ and then examining the response in resulting current $j(t)$. For the complex frequency-dependent conductivity, Smith used the following Fourier transform $j(t)$:

$$\tilde{\sigma}(\omega)=\int_0^\infty j(t)\exp(i\omega t)dt\,. \tag{13.12}$$

A field impulse, which exceeds every other force acting on the system, allows to assume the hypothesis that initially the electrons can be considered totally free; therefore, it holds:

$$j(0)=n^*e^2/m\,, \tag{13.13}$$

with $n^*$ effective electron density. Being $j(t)$ and $\tilde{\sigma}(\omega)$ Fourier transforms of each other, by the rule of the sum of the oscillator strength on $\tilde{\sigma}(\omega)$, its real part results:

$$\int_0^\infty \mathrm{Re}\,\tilde{\sigma}(\omega)d\omega=\frac{\pi}{2}j(0)=\frac{\omega_p^2}{8}, \tag{13.14}$$

with $\omega_p=\sqrt{n^*e^2/(\varepsilon_0 m)}$ plasma frequency. If the initial current decays exponentially to its initial value with relaxation time $\tau$, it is:

$$j(t)/j(0)=\exp(-t/\tau), \tag{13.15}$$

from which we get the standard Drude formula:

$$\tilde{\sigma}(\omega)=(n^*e^2\tau/m)/(1-i\omega\tau)\,. \tag{13.16}$$

Equation 13.16 can be considered as the first term of a series of the form:

$$j(t)/j(0)=\exp(-t/\tau)\left[1+\sum_{n=1}^{\infty}c_n\,(t/\tau)^n/n!\right]. \tag{13.17}$$

The $c_n$ factor holds into account of the original electrons speed remained after the $n$-th collision. The analytical form of the complex conductivity is:

$$\tilde{\sigma}(\omega)=\frac{n^*e^2\tau}{m(1-i\omega\tau)}\left[1+\sum_{n=1}^{\infty}\frac{c_n}{(1-i\omega\tau)^n}\right]. \tag{13.18}$$

This formula generalises that of Drude through factors $c_n$.

## 13.6 Linear Response Theory with a New Idea

Let us consider a system with a Hamiltonian of the form:

$$H=H_0+H_1, \tag{13.19}$$

with $H_1$ having small effects with respect to $H_0$; we write $H_1$ as:

$$H_1=\lambda A e^{-i\omega t}e^{\eta t}, \tag{13.20}$$

With $\lambda$ a real quantity and $\eta$ positive, so that in remote past the perturbation is negligible:

Adiabatic representation:

$$\lim_{t\to-\infty} H_1=0. \tag{13.21}$$

In the case of an electric field of frequency $\omega$, it is

$$H_1=e\mathbf{E}\cdot\mathbf{r}. \tag{13.22}$$

If the electric field is constant in space and its evolution depends on time, we can write it as:

$$\mathbf{E}=\mathbf{E}_0 e^{i\omega t}. \tag{13.23}$$

The time dependent corresponding current is:

$$\mathbf{J}(t)=\sigma(\omega)\mathbf{E}(t). \tag{13.24}$$

The conductivity $\sigma(\omega)$ is, in general, a complex function of the frequency $\omega$ and can be deduced by linear response theory as:

$$\sigma_{\beta\alpha}(\omega)=\frac{e^2}{\hbar V}\int_0^{\infty}dt\, e^{-i\omega t}\int_0^{\beta}d\lambda<\mathbf{v}^{\alpha}(t-i\lambda)\mathbf{v}^{\beta}(0)>, \tag{13.25}$$

where $\beta = 1/k_BT$, $V$ is the volume of the system, $T$ its temperature and $v^{\alpha}(t)$ the $\alpha$ component of velocity of charge carrier at time $t$.

It is possible to get the velocities correlation function $<\boldsymbol{v}^{\alpha}(0)\, \boldsymbol{v}^{\beta}(t)>_T$ by inversion of Eq. 13.25. However, the presence of an integration from 0 to +∞ in it makes no easy such procedure. By modifying Eq. 13.25 after consideration of the entire time axis, it is possible, however, to make a complete Fourier transform. If we consider the real part of the conductivity, such an extension to the entire time axis is viable. Considering valid results at high temperatures, the condition $\hbar\omega/k_B T<<1$ is also verified.

Following the standard time-dependent approach, we can get a general formula for the linear response of a dipole moment density $\boldsymbol{B} = e\boldsymbol{r}/V$ in the $\beta$ direction with the electric field $\boldsymbol{E}$ in the $\alpha$ direction, where $V$ is the volume of the system. This allows to deduce the susceptibility $\chi(\omega)$, that is correlated to $\sigma(\omega)$ through the relation:

$$1+4\pi\,\chi(\omega)=1+4\pi i\frac{\sigma(\omega)}{\omega}. \tag{13.26}$$

After analytical calculation, the real part of the complex conductivity $\sigma(\omega)$ appears as:

$$\mathrm{Re}\,\sigma_{\beta\alpha}(\omega)=\frac{e^2\,\omega\pi}{V\hbar}S_{\beta\alpha}(\omega)\left(1-e^{-\hbar\omega/KT}\right), \tag{13.27}$$

with

$$S_{\beta\alpha}(\omega)=\int_{-\infty}^{+\infty}dt\left\langle \boldsymbol{r}^{\alpha}(0)\,\boldsymbol{r}^{\beta}(t)\right\rangle_T e^{-i\omega t}. \tag{13.28}$$

The part $\langle.....\rangle_T$ in Eq. 13.28 is the thermal average and the exponential factor arises from equilibrium thermal weights for Fermi particles. Considering the identity:

$$v=\frac{d}{dt}r=\frac{i}{\hbar}[H,r], \tag{13.29}$$

Equation 13.28 can be written in a form containing the velocities correlation function instead of the position correlation function. We obtain:

$$\mathrm{Re}\,\sigma_{\beta\alpha}(\omega)=\frac{e^2}{2VKT}\int_{-\infty}^{+\infty}dt\left\langle \boldsymbol{v}^{\alpha}(0)\boldsymbol{v}^{\beta}(t)\right\rangle_T e^{-i\omega t}. \tag{13.30}$$

The integral in Eq. 13.30 spans the entire $t$-axis, so we can perform the complete inverse Fourier transform of this equation, obtaining:

$$<\mathbf{v}^{\alpha}(0)\,\mathbf{v}^{\beta}(t)>_T = \frac{KTV}{\pi e^2}\int_{-\infty}^{+\infty} d\omega\, \mathrm{Re}\,\sigma_{\beta\alpha}(\omega)e^{i\omega t}. \tag{13.31}$$

The *new idea* is, therefore, the possibility to perform a complete inversion of Eq. 13.30 on temporal scale, i.e., considering the entire time axis (−∞, +∞), not the half-time axis (0, +∞), as usually considered in the literature, possible if we consider the real part of the complex conductivity $\sigma(\omega)$.

Considering contour integration with the residue theorem in Eq. 13.31, the integral is determined by the poles of Re $\sigma(\omega)$. This leads to an exact formulation and gives a powerful method to get the analytical form of the velocities correlation function, the mean square deviation of position and the diffusion coefficient.

The Fourier transforms considered in the calculation are as follows:

$$\tilde{f}(\omega) = \frac{1}{2\pi}\int_{-\infty}^{+\infty} dt\, f(t)e^{-i\omega t}\ ; \tag{13.32}$$

$$f(t) = \int_{-\infty}^{+\infty} d\omega\, \tilde{f}(\omega)e^{i\omega t}. \tag{13.33}$$

The inversion of relation of Eq. 13.25 extending the integration to the entire time axis (−∞, +∞) is one of the key ideas of this new formulation, called Di Sia model (DS model). Its main advantage is the analytical derivation, avoiding numerical equations for the description of the dynamic behaviour of nanosystems; these relations are functions of parameters that are experimentally obtainable through time-resolved techniques.

A significant consequence of the application of such a model is the mathematical justification of the unexpected experimental results of high initial mobility of charge carriers in semiconductor devices based on nanomaterials, such as dye sensitised solar cells (DSSCs).

The increase of the diffusion coefficient leads to a raise in efficiency of such electrochemical cells. The new model is able to describe quite adequately the transport properties of nanobiomaterials.

## 13.7 Other Key Functions

Besides the velocities correlation function, the second interesting quantity is the mean square deviation of position of particles at equilibrium $R^2(t)$, defined as:

$$R^2(t) = \left\langle [\boldsymbol{R}(t) - \boldsymbol{R}(0)]^2 \right\rangle, \tag{13.34}$$

where $R(t)$ is the position vector at time $t$. Considering that it holds:

$$\boldsymbol{R}(t) - \boldsymbol{R}(0) = \int_0^t \boldsymbol{v}(t')dt', \tag{13.35}$$

where $\boldsymbol{v}(t)$ is the velocity and considering that for a homogeneous system in equilibrium, the invariance property for time translation holds, it is possible to rewrite Eq. 13.34 by means of a transformation of coordinates related to the region of integration. We get:

$$R^2(t) = 2\int_0^t dt'(t - t')\langle \boldsymbol{v}(t')\cdot\boldsymbol{v}(0)\rangle. \tag{13.36}$$

The mean square deviation of position can therefore be evaluated through the velocities correlation function. Thanks to Eq. 13.31, it is also possible to write the diffusion coefficient $D$ in analytical form:

$$D(t) = \frac{1}{2}\frac{d}{dt}R^2(t) = \int_0^t dt'\langle \boldsymbol{v}(t')\cdot\boldsymbol{v}(0)\rangle. \tag{13.37}$$

Relations 13.31, 13.36 and 13.37 represent a complete analytical description of the most important parameters concerning transport phenomenon.

## 13.8 The Complex Conductivity σ(ω)

Let us consider a particle in a region in which there is an electric time-oscillating field, directed for example along $z$-axis, with an elastic-type force acting on the system; the dynamic equation of the particle can be written as:

$$m\ddot{z} = -kz + eE_o e^{-i\omega t}, \tag{13.38}$$

where $m$ is the mass of the particle and $k$ is the strength constant of the oscillators. By Eq. 13.38, we can get the *diffusion equation* of the harmonic oscillator:

$$\ddot{z} + \omega_0^2 z = \frac{eE_0}{m} e^{-i\omega t} , \qquad (13.39)$$

with $\omega_0 = \sqrt{k/m}$ . Its solution is given by the general solution of the associated homogeneous equation added to a particular solution. The homogeneous solution can be neglected because it does not depend from the applied electric field and would lead to a zero current; only the part dependent by the field does a contribution. Therefore, we take as general solution the following:

$$z(t) = z_0 e^{-i\omega t} . \qquad (13.40)$$

An adequate generalisation is obtained by adding a term that takes into account the velocity (friction-type term), in the form – $\lambda z$, with $\lambda = m/\tau$ and $\tau$ relaxation time. The $z_0$ value is:

$$z_0 = \frac{eE_0}{m} \frac{1}{\omega_0^2 - \omega^2 - i\lambda\omega} . \qquad (13.41)$$

Considering the current density in the field direction:

$$j_z = ne\dot{z} = nez_0(-i\omega)e^{i\omega t} = \sigma E , \qquad (13.42)$$

we can get the complex conductivity as a function of frequency:

$$\sigma = \frac{i\omega ne^2}{m(\omega^2 - \omega_0^2 + i\lambda\omega)} . \qquad (13.43)$$

The real part of the complex conductivity (Eq. 13.43) is:

$$\mathrm{Re}\,\sigma(\omega) = \frac{\tau ne^2}{m} \frac{\omega^2/\tau^2}{(\omega_0^2 - \omega^2)^2 + \omega^2/\tau^2} . \qquad (13.44)$$

## 13.9 Behaviour of Complex Conductivity σ(ω)

The behaviour of real and imaginary parts of the complex conductivity $\sigma(\omega)$ is consistent with what is required by the Kramers–Krönig relations:

$$\sigma_1(\omega) = \frac{1}{\pi} \wp \int_{-\infty}^{+\infty} d\omega' \frac{\sigma_2(\omega')}{\omega' - \omega} , \qquad (13.45)$$

$$\sigma_2(\omega) = -\frac{1}{\pi}\wp\int_{-\infty}^{+\infty} d\omega' \frac{\sigma_1(\omega')}{\omega' - \omega}, \tag{13.46}$$

with $\sigma_1(\omega)$ and $\sigma_2(\omega)$ real and imaginary part of $\sigma(\omega)$, respectively, and $\wp$ Cauchy's principal value of the integral.

These relations hold as result of the causal behaviour of the system and this leads to the analyticity of function $\sigma(\omega)$ in the complex plane. Consequently, the velocities correlation function (Eq. 13.31) can be exactly evaluated by means of the residues theorem.

## 13.10 The Poles of Conductivity $\sigma(\omega)$

Considering the DS model, it is useful to split the case $\omega_0 = 0$ from the case $\omega_0 \neq 0$; for the latter three cases are possible, following the sign of the quantity $\Delta = 4\tau^2\omega_0^2 - 1$.

(a) $\omega_0 = 0$ implies that Eq. 13.44 becomes:

$$\mathrm{Re}\,\sigma(\omega) = \frac{\tau n e^2}{m}\frac{1}{1+\tau^2\omega^2}. \tag{13.47}$$

The respective poles are at frequencies:

$$\omega_{1/2} = \pm\frac{i}{\tau}. \tag{13.48}$$

The residue for integration in the complex plane is:

$$Res\left(\mathrm{Re}\,\sigma\left(\frac{i}{\tau}\right)\right) = -\frac{\tau}{2} i e^{-\frac{t}{\tau}}, \tag{13.49}$$

having considered positive times.

(b) $\omega_0 \neq 0$ implies the poles:

$$\omega_{1/2} = \frac{1}{2\tau}\left(i \pm \sqrt{\Delta}\right). \tag{13.50}$$

It is possible to write:

(i) $\Delta > 0 \Rightarrow \omega_{1/2} = \frac{1}{2\tau}(i \pm \alpha_R)$, (13.51)

with $\alpha_R = \sqrt{4\tau^2\omega_0^2 - 1}$. (13.52)

(ii) $\Delta = 0 \Rightarrow \omega_{1/2} = \frac{1}{2\tau} i$ . (13.53)

(iii) $\Delta < 0 \Rightarrow \omega_{1/2} = \frac{i}{2\tau}(1 \pm \alpha_I)$, (13.54)

with $\alpha_I = \sqrt{1 - 4\tau^2 \omega_0^2}$ . (13.55)

## 13.11 Premises of Quantum and Relativistic Versions of DS Model

In general reasoning and application of the DS model, it is considered that the transition between classical and quantum regime occurs at the nanometer length scale. In fact, if one considers the de Broglie wavelength of thermal electrons and looks at its comparability to the length of the system, such lengths are comparable when they are of the order of nanometer.

The model allows a generalisation of the classical Drude–Lorentz formula based on linear response theory and of the resonant plasmonic mode; in addition to the classic treatment, it allows a detailed description at a quantum level for the dynamic response of carriers in nanoscale systems.

(a) **Quantum case:** Let us consider the effect of a frequency-dependent electric field of the form as given in Eq. 13.23. We follow the time-dependent perturbation theory, treating the factor $e\,\boldsymbol{E} \cdot \boldsymbol{r}$ as a perturbing potential (in this context $\boldsymbol{r}$ is the position vector of the particle).

The matrix elements of the dipole moment of electron in the direction of the electric field, between the initial $\Phi_0$ and the excited $\Phi_j$ states, are given by:

$$e x_{j0} = \int \Phi_j^* e x \Phi_0 \, dr \, . \tag{13.56}$$

Defining the oscillator strength of the $j$-th transition as:

$$f_j = \frac{2m}{\hbar^2} \sum_j \hbar \omega_j \left| x_{0j} \right|^2 , \tag{13.57}$$

and considering the relation between permittivity and conductivity of the system:

$$\varepsilon(\omega) = 1 + 4\pi i \frac{\sigma(\omega)}{\omega} = 1 + \frac{4\pi N e^2}{m} \sum_j \frac{f_j}{\omega_j^2 - \omega^2 + i\omega\Gamma_j}, \tag{13.58}$$

we obtain the relation:

$$\frac{i\sigma(\omega)}{\omega} = \frac{1}{4\pi} \sum_i \frac{\omega_{p_i}^2}{(\omega_i^2 - \omega^2) - i\omega\Gamma_i}, \tag{13.59}$$

with:

$$\omega_{p_i}{}^2 = \frac{4\pi N e^2}{m} f_i\,. \tag{13.60}$$

From Eqs. 13.58, 13.59 and 13.60, we have:

(i) $\omega_i = (E_i - E_0)/\hbar$,

(ii) $E_i$, $E_0$ energies of the excited and the ground states, respectively,

(iii) $\Gamma_i = 1/\tau_i$ inverse of the decay time of every mode,

(iv) $N$ density of carriers.

As in the classical case, the analytical calculation formally brings to Eq. 13.31, but with the fundamental difference that the real part of conductivity inside the integral is calculated at the quantum level. The key factors incorporating the quantum behaviour are the weights $f_i$. The relaxation times can be obtained by $\tau_i = 1/\Gamma_i$ and weights $f_i$ by Eq. 13.60. The calculation of $N$ can be exactly obtained by Eq. 13.60, considering that it holds $\sum_i f_i = 1$.

(b) **Relativistic case:** Let us consider possible relativistic velocities of carriers (not necessarily electrons) inside a nanostructure. We consider as focus point the variation of mass of the carrier. Starting with the dynamics law:

$$\frac{d}{dt}(m_{\text{part}}\boldsymbol{v}) = \sum_i \boldsymbol{F}_i\,, \tag{13.61}$$

we studied the condition of relativistic variation of the mass along an $x$-axis in the fixed ground reference frame. About the forces acting on the generic carrier, it has been considered an outer passive elastic-type force of the form $F_{el} = Kx$, a passive friction-type force of the form $F_{fr} = \lambda\dot{x}$, depending by velocity and with $\lambda = m_{part}/\tau$ and an outer oscillating electric field $E = e\,E_0\,e^{-i\omega t}$, considering solutions of the form $x = x_0\,e^{-i\omega t}$. The analytical calculation brings to three sets of results, in relation to the value of the quantity $\Delta$, as in the classical and quantum cases.

## 13.12 Classical Results

Using relations Eqs. 13.31, 13.36 and 13.37, we obtained the following results:

(a) **Case $\Delta > 0$**

$$\langle v(0)\cdot v(t)\rangle = \left(\frac{k_B T}{m^*}\right)\exp\left(-\frac{t}{2\tau}\right)\left[\cos\left(\frac{\alpha_R}{2}\frac{t}{\tau}\right)-\frac{1}{\alpha_R}\sin\left(\frac{\alpha_R}{2}\frac{t}{\tau}\right)\right], \tag{13.62}$$

$$R^2(t) = \left(\frac{k_B T}{m^*}\frac{2}{\omega_0^2}\right)\left[-\frac{1}{\alpha_R}\sin\left(\frac{\alpha_R}{2}\frac{t}{\tau}\right)\exp\left(-\frac{t}{2\tau}\right)\right. \\ \left. -\cos\left(\frac{\alpha_R}{2}\frac{t}{\tau}\right)\exp\left(-\frac{t}{2\,\tau}\right)+1\right], \tag{13.63}$$

$$D(t) = 2\left(\frac{k_B T}{m^*}\right)\left(\frac{\tau}{\alpha_R}\right)\left[\sin\left(\frac{\alpha_R}{2}\frac{t}{\tau}\right)\exp\left(-\frac{t}{2\tau}\right)\right], \tag{13.64}$$

with $\alpha_R^2 = 4\tau^2\omega_0^2 - 1$. (13.65)

(b) **Case $\Delta < 0$**

$$\langle v(0)\cdot v(t)\rangle = \frac{1}{2}\left(\frac{k_B T}{m^*}\right)\left(\frac{1}{\alpha_I}\right)\left[(1+\alpha_I)\exp\left(-\frac{(1+\alpha_I)}{2}\frac{t}{\tau}\right)\right. \\ \left. -(1-\alpha_I)\exp\left(-\frac{(1-\alpha_I)}{2}\frac{t}{\tau}\right)\right], \tag{13.66}$$

$$R^2(t) = 4\left(\frac{k_BT}{m^*}\right)(\tau^2)\left[\frac{1}{\alpha_I(1+\alpha_I)}\exp\left(-\frac{(1+\alpha_I)}{2}\frac{t}{\tau}\right)\right.$$
$$\left. -\frac{1}{\alpha_I(1-\alpha_I)}\exp\left(-\frac{(1-\alpha_I)}{2}\frac{t}{\tau}\right)+\frac{2}{1-\alpha_I{}^2}\right], \quad (13.67)$$

$$D(t) = \left(\frac{k_BT}{m^*}\right)(\tau)\left(\frac{1}{\alpha_I}\right)\left[-\exp\left(-\frac{1+\alpha_I}{2}\frac{t}{\tau}\right)+\exp\left(-\frac{1-\alpha_I}{2}\frac{t}{\tau}\right)\right], \quad (13.68)$$

with $\alpha_I^2 = 1 - 4\tau^2\omega_0^2$. (13.69)

(c) **Case $\omega_0 = 0$**

$$\langle \mathbf{v}(0)\cdot\mathbf{v}(t)\rangle = \left(\frac{k_BT}{m^*}\right)\exp\left(-\frac{t}{\tau}\right), \quad (13.70)$$

$$R^2(t) = 2\left(\frac{k_BT}{m^*}\right)(\tau^2)\left(\exp\left(-\frac{t}{\tau}\right)+\frac{t}{\tau}-1\right), \quad (13.71)$$

$$D(t) = \left(\frac{k_BT}{m^*}\right)(\tau)\left[-\exp\left(-\frac{t}{\tau}\right)+1\right]. \quad (13.72)$$

## 13.13 Quantum Results

For quantum case, we have:

(a) **Case $\Delta > 0$**

$$\langle \mathbf{v}(0)\cdot\mathbf{v}(t)\rangle = \left(\frac{k_BT}{m^*}\right)$$
$$\times\sum_i\left(f_i\exp\left(-\frac{t}{2\tau_i}\right)\left[\cos\left(\frac{\alpha_{iR}}{2}\frac{t}{\tau_i}\right)-\frac{1}{\alpha_{iR}}\sin\left(\frac{\alpha_{iR}}{2}\frac{t}{\tau_i}\right)\right]\right), \quad (13.73)$$

$$R^2(t) = 2\left(\frac{k_BT}{m^*}\right)\sum_i\left[\frac{f_i}{\omega_i^2}\left\{-\frac{1}{\alpha_{iR}}\sin\left(\frac{\alpha_{iR}}{2}\frac{t}{\tau_i}\right)\times\exp\left(-\frac{t}{2\tau_i}\right)\right.\right.$$
$$\left.\left.-\cos\left(\frac{\alpha_{iR}}{2}\frac{t}{\tau_i}\right)\exp\left(-\frac{t}{2\tau_i}\right)+1\right\}\right], \quad (13.74)$$

$$D(t)=2\left(\frac{k_{\mathrm{B}}T}{m^*}\right)\sum_i\left(\frac{f_i\tau_i}{\alpha_{i\mathrm{R}}}\sin\left(\frac{\alpha_{i\mathrm{R}}}{2}\frac{t}{\tau_i}\right)\exp\left(-\frac{t}{2\tau_i}\right)\right), \tag{13.75}$$

with $\alpha_{i\mathrm{R}}^2=4\tau_i^2\omega_i^2-1$. (13.76)

(b) **Case Δ < 0**

$$\langle \boldsymbol{v}(0)\cdot\boldsymbol{v}(t)\rangle=\frac{1}{2}\left(\frac{k_{\mathrm{B}}T}{m^*}\right)\sum_i\left[\left(\frac{f_i}{\alpha_{i\,I}}\right)\left\{(1+\alpha_{i\,I})\exp\left(-\frac{(1+\alpha_{i\,I})}{2}\frac{t}{\tau_i}\right)\right.\right.$$
$$\left.\left.-(1-\alpha_{i\,I})\exp\left(-\frac{(1-\alpha_{i\,I})}{2}\frac{t}{\tau_i}\right)\right\}\right], \tag{13.77}$$

$$R^2(t)=4\left(\frac{k_{\mathrm{B}}T}{m^*}\right)\sum_i\left[f_i\tau_i^2\left\{\frac{1}{\alpha_{i\,I}(1+\alpha_{i\,I})}\exp\left(-\frac{(1+\alpha_{i\,I})}{2}\frac{t}{\tau_i}\right)\right.\right.$$
$$\left.\left.-\frac{1}{\alpha_{i\,I}(1-\alpha_{i\,I})}\exp\left(-\frac{(1-\alpha_{i\,I})}{2}\frac{t}{\tau_i}\right)+\frac{2}{1-\alpha_{i\,I}^2}\right\}\right], \tag{13.78}$$

$$D(t)=\left(\frac{k_{\mathrm{B}}T}{m^*}\right)\sum_i\left[\left(\frac{f_i\tau_i}{\alpha_{i\,I}}\right)\left\{-\exp\left(-\frac{1+\alpha_{i\,I}}{2}\frac{t}{\tau_i}\right)\right.\right.$$
$$\left.\left.+\exp\left(-\frac{1-\alpha_{i\,I}}{2}\frac{t}{\tau_i}\right)\right\}\right], \tag{13.79}$$

with $\alpha_{i\mathrm{I}}^2=1-4\tau_i^2\omega_i^2$. (13.80)

(c) **Case $\omega_0 = 0$**

$$\langle \boldsymbol{v}(0)\cdot\boldsymbol{v}(t)\rangle=\left(\frac{k_{\mathrm{B}}T}{m^*}\right)f_0\exp\left(-\frac{t}{\tau_0}\right), \tag{13.81}$$

$$R^2(t)=2\left(\frac{k_{\mathrm{B}}T}{m^*}\right)\sum_i\left(f_i\tau_i^2\left(\exp\left(-\frac{t}{\tau_i}\right)+\frac{t}{\tau_i}-1\right)\right), \tag{13.82}$$

$$D(t)=\left(\frac{k_{\mathrm{B}}T}{m^*}\right)\sum_i\left(f_i\tau_i\left(-\exp\left(-\frac{t}{\tau_i}\right)+1\right)\right). \tag{13.83}$$

($m^*$ = effective mass)

## 13.14 Relativistic Results

For relativistic case, we have:

(a) **Case $\Delta_{\text{rel}} > 0$**

$$\langle \boldsymbol{v}(0)\cdot\boldsymbol{v}(t)\rangle = \left(\frac{k_B T}{m_0}\right)\left(\frac{1}{\gamma\rho}\right)\exp\left(-\frac{t}{2\tau\rho}\right)$$
$$\times\left[\cos\left(\frac{\alpha_{R_{\text{rel}}}}{2\rho}\frac{t}{\tau}\right) - \frac{1}{\alpha_{R_{\text{rel}}}}\sin\left(\frac{\alpha_{R_{\text{rel}}}}{2\rho}\frac{t}{\tau}\right)\right], \quad (13.84)$$

$$R^2(t) = 2\left(\frac{k_B T}{m_0}\right)\left(\frac{1}{\omega_0^2}\right)$$
$$\left[-\frac{1}{\alpha_{R_{\text{rel}}}}\sin\left(\frac{\alpha_{R_{\text{rel}}}}{2\rho}\frac{t}{\tau}\right)\exp\left(-\frac{t}{2\tau\rho}\right) - \cos\left(\frac{\alpha_{R_{\text{rel}}}}{2\rho}\frac{t}{\tau}\right)\exp\left(-\frac{t}{2\tau\rho}\right) + 1\right],$$
$$(13.85)$$

$$D(t) = 2\left(\frac{k_B T}{m_0}\right)\left(\frac{1}{\gamma}\right)\left(\frac{\tau}{\alpha_{R_{\text{rel}}}}\right)\left[\sin\left(\frac{\alpha_{R_{\text{rel}}}}{2\rho}\frac{t}{\tau}\right)\exp\left(-\frac{t}{2\rho\tau}\right)\right],$$
$$(13.86)$$

with $\alpha_{R_{\text{rel}}}^2 = 4\gamma\tau^2\omega_0^2 - 1$. (13.87)

(b) **Case $\Delta_{\text{rel}} < 0$**

$$\langle \boldsymbol{v}(0)\cdot\boldsymbol{v}(t)\rangle = \frac{1}{2}\left(\frac{k_B T}{m_0}\right)\left(\frac{1}{\gamma\rho}\right)\left(\frac{1}{\alpha_{I_{\text{rel}}}}\right)\left[(1+\alpha_{I_{\text{rel}}})\exp\left(-\frac{(1+\alpha_{I_{\text{rel}}})}{2\rho}\frac{t}{\tau}\right)\right.$$
$$\left. -(1-\alpha_{I_{\text{rel}}})\exp\left(-\frac{(1-\alpha_{I_{\text{rel}}})}{2\rho}\frac{t}{\tau}\right)\right], \quad (13.88)$$

$$R^2(t) = 4\left(\frac{k_B T}{m_0}\right)(\tau^2\gamma)\left[\frac{1}{\alpha_{I_{\text{rel}}}(1+\alpha_{I_{\text{rel}}})}\exp\left(-\frac{(1+\alpha_{I_{\text{rel}}})}{2\rho}\frac{t}{\tau}\right)\right.$$
$$\left. -\frac{1}{\alpha_{I_{\text{rel}}}(1-\alpha_{I_{\text{rel}}})}\exp\left(-\frac{(1-\alpha_{I_{\text{rel}}})}{2\rho}\frac{t}{\tau}\right) + \frac{2}{(1-\alpha_{I_{\text{rel}}}^2)}\right], \quad (13.89)$$

$$D(t)=\left(\frac{k_B T}{m_0}\right)\left(\frac{\tau}{\gamma}\right)\left(\frac{1}{\alpha_{I_{\text{rel}}}}\right)\left[-\exp\left(-\frac{1+\alpha_{I_{\text{rel}}}}{2\rho}\frac{t}{\tau}\right)+\exp\left(-\frac{1-\alpha_{I_{\text{rel}}}}{2\rho}\frac{t}{\tau}\right)\right], \tag{13.90}$$

with $\alpha_{I_{\text{rel}}}^2 = 1-4\gamma\tau^2\omega_0^2$. (13.91)

(c) **Case $\omega_0 = 0$:** It reduces to relativistic Drude model.

($m_0$ = rest mass)

It is also: $\alpha_{I_{\text{rel}}} = \sqrt{\Delta_{I_{\text{rel}}}}$, $\alpha_{R_{\text{rel}}} = \sqrt{\Delta_{R_{\text{rel}}}}$, $\gamma = 1/\sqrt{1-\beta^2}$, $\beta = v/c$, $\rho = \gamma^2$.

Equations, governed by parameter of type $\alpha_I$, are a superposition of exponentials; the behaviour of curves is similar to a typical Drude–Lorentz behaviour.

Equations, governed by parameter of type $\alpha_R$, are a product of an exponential with a sinusoidal type function; the behaviour of curves is a typical damped oscillation in time.

The model contains also a gauge factor, which allows its use from sub-pico-level to macro-level. Interesting applications have been performed for economics, neuro-science, brain processes and nano-medicine.

Acting on all chemical, physical, structural and model-intrinsic parameters, i.e.,

(i) The temperature $T$ of the system,

(ii) The parameters $\alpha_I$ and $\alpha_R$,

(iii) The values of $\tau_i$ and $\omega_i$,

(iv) The effective mass $m^*$,

(v) Variations of the chiral vector,

(vi) The quantum weights of each mode in the quantum case,

(vii) The carrier density $N$,

(viii) The velocity of carriers,

it is possible to perform a fine and accurate tuning of $<\mathbf{v}(t)\cdot\mathbf{v}(0)>_T$, $R^2(t)$ and $D(t)$ and therefore to calibrate the performance of nano-bio-devices and an 'a-priori' and 'a-posteriori' application of the model.

## 13.15 Examples of Application

It has been studied with the DS model the currently most common materials used in nanotechnology, i.e., zinc oxide (ZnO), titanium dioxide ($TiO_2$), gallium arsenide (GaAs), silicon (Si), carbon nanotubes (CN) and also 'last generation' materials, like cadmium telluride (CdTe), cadmium sulphide (CdS), copper indium selenide (CIS), copper indium gallium selenide (CIGS). It follows various application examples:

(a) Let us consider a $TiO_2$ nanostructure. At temperature $T = 300$ K, considering that the effective mass is $m^* = 6\, m_e$, with $m_e$ the mass of the electron, the behaviour of the velocities correlation function versus $t/\tau$ for two values of the parameter $\alpha_R$ ($\Delta > 0$) ($\alpha_{R1} = 10$; $\alpha_{R2} = 20$) is as shown in Fig. 13.1. It is $\tau = 0.1 \times 10^{-13}$ s. We note the classical damped oscillating behaviour.

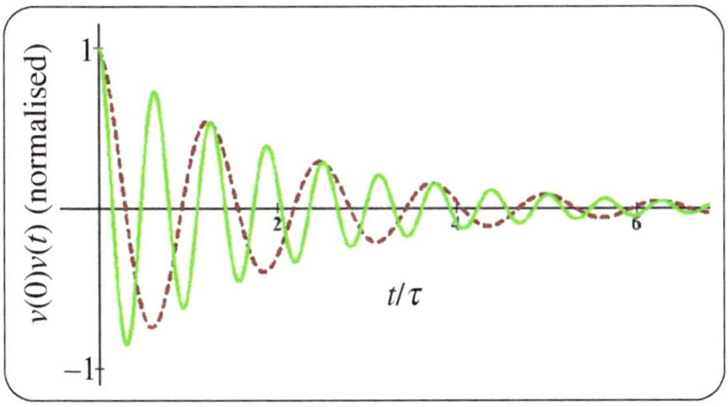


**Figure 13.1** Behaviour of the velocities correlation function versus time with the parameter $\alpha_R$.

(b) Using the numbers of the previous example, but in this case, considering the parameter $\alpha_I$ ($\Delta < 0$) ($\alpha_{I1} = 0.1$; $\alpha_{I2} = 0.5$; $\alpha_{I3} = 0.9$), we get the behaviour shown in Fig. 13.2.

For $\alpha_i \to 0$, curves reproduce the results found using terahertz spectroscopy in studying nanomaterials with Drude–Smith models for ZnO nanowires, polycrystalline and nanoparticle films, as well as dye-sensitised nanocrystalline colloidal $TiO_2$ films. Excluding the Drude case ($\omega_0 = 0$), the velocities correlation function is not a single decreasing exponential of time, but a more complicated combination of exponentials, or an oscillating function of time, as technically shown in previous paragraphs. The results give a precise indication on response times of a system subjected to charge motion. It has been verified that the two cases $\Delta > 0$ and $\Delta < 0$ occur in such systems.

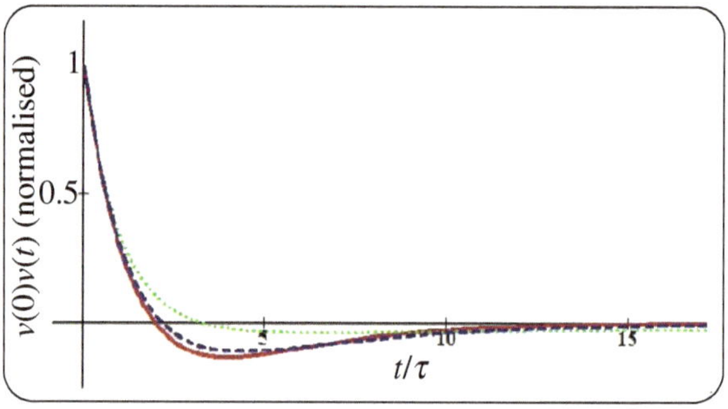


**Figure 13.2** Behaviour of the velocities correlation function versus time with the parameter $\alpha_l$.

(c) In Fig. 13.3, the velocities correlation function has been compared with the fit of experimental data using $TiO_2$, but similar results are obtainable in the case of films and nanostructures of ZnO, so as considering different materials.

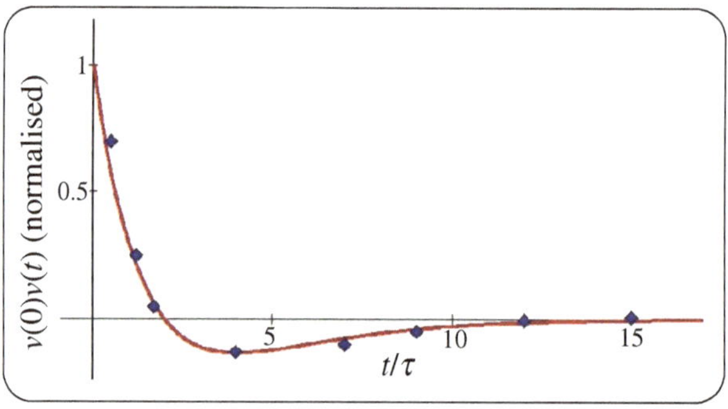


**Figure 13.3** Dots represent experimental data; the curve derives by the DS model. Error bars are not present for simplicity.

The results indicate that there is a positive velocities correlation in less time than the relaxation time, while at later times (comparable with those of relaxation) the velocity changes orientation. At longer times, the correlation tends to zero, leading to a negligible current in the system. This feature is also predicted by the Smith model, but in DS model it appears without assumptions similar to the transport parameter $c_n$ of the Smith model.

(d) In relation to $R^2$, from which we can get the effective space travelled by charges, at large times the Drude–Lorentz term leads to an $R^2$ approaching a constant value (Fig. 13.4), while the Drude term alone is the dominant one.

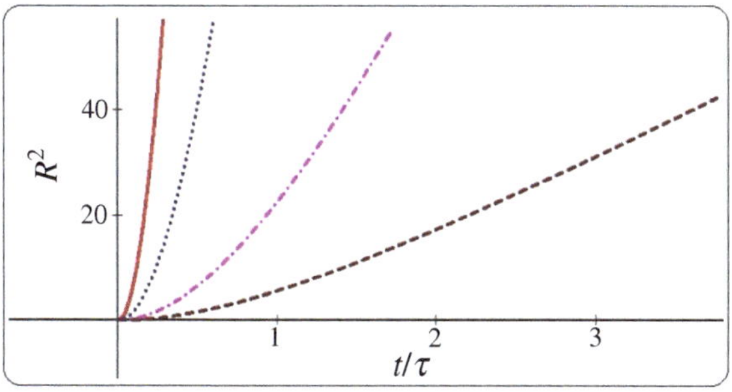


**Figure 13.4** Behaviour of $R^2$ versus time (see text for details).

Therefore, for sufficiently large times, only the Drude term survives. The linear relation at large times becomes quadratic at smaller times. The cross-over between the two regimes occurs at times comparable to the scattering time. This means that diffusion occurs after sufficient time has elapsed, so that scattering events become significant, while at smaller times the motion is essentially ballistic. In Fig. 13.4 we show $R^2$ ($\times 10^{-14}$ cm$^2$) versus $t/\tau$ for some representative values of $\tau$ (of the order of $10^{-13}$–$10^{-14}$ s), typical of doped silicon ($T$ = 300 K).

(e) In Fig. 13.5 we have $R^2$ ($\times 10^{-14}$ cm$^2$) versus $t/\tau$ for two values of $\tau$ ($\omega_0 = 1.12 \times 10^{11}$ Hz dashed line; $\omega_0 = 2.24 \times 10^{11}$ Hz solid line) for $TiO_2$ nanoparticle films ($m^* = 6m_e$, $T$ = 300 K), considering large times $x = t/\tau >> 1$. We observe that curves reach a plateau value at sufficiently long times and that the slope within a given time interval increases with $\tau$.

As consequences of this behaviour, the plateau of $R^2$ may become larger than the size of the nanoparticles composing the films, depending on the parameters $\omega_0$ and $\tau$, indicating that carriers created at time $t = 0$ have enhanced mobility in the nanoporous films at small times, on the order of a few $\tau$, in contrast with a commonly expected low mobility in the disordered $TiO_2$ network.

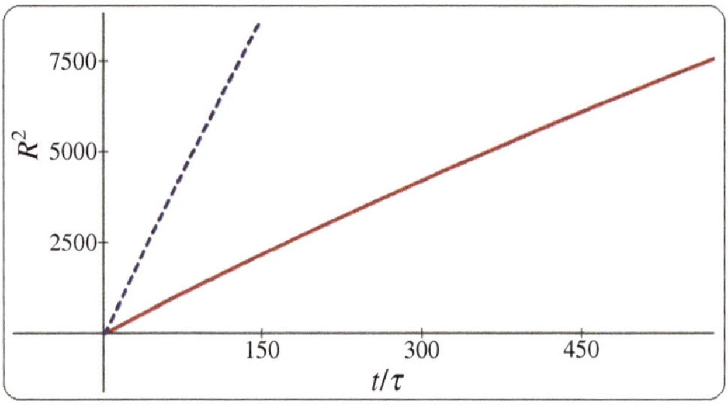


**Figure 13.5** Behaviour of $R^2$ versus time (see text for details).

(f) Figure 13.6 represents $R^2$ ($\times 10^{-14}$ cm$^2$) versus $t/\tau$ for two values of $\tau$ ($\omega_0 = 0.5 \times 10^{13}$ Hz solid line; $\omega_0 = 10^{13}$ Hz dashed line) for $TiO_2$ ($m^* = 6\ m_e$, $T = 300$ K). This case corresponds to short time behaviour and large frequencies, with the scattering time in the range $10^{-13}$–$10^{-14}$ s.

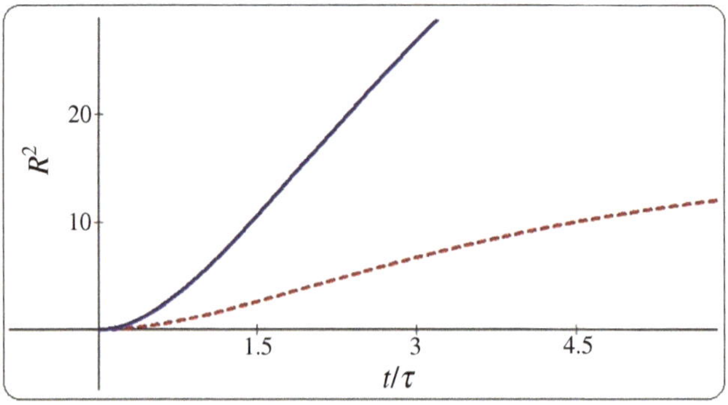


**Figure 13.6** $R^2$ versus time (see text for details).

(g) Figure 13.7 shows the diffusion coefficient $D$ as a function of time considering a ZnO nanostructure, with a fixed relaxation time and for different values of the parameter $\alpha_l$ running in the range (0,1), with $\tau \cong 0.7 \times 10^{-13}$ s.

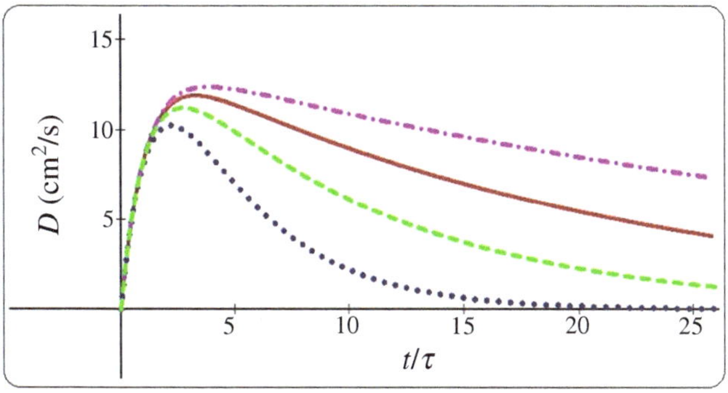


**Figure 13.7** Diffusion versus time (see text for details).

(h) In Fig. 13.8 we see the behaviour of the diffusion coefficient $D$ for $\alpha_l$ varying in the interval (0.1–1) with dots representing experimental data of GaAs photoconductivity with respective error bars. Dots are derived by THz spectroscopy, the error bars on dots represent the distribution of the experimental data and the lines derive by DS model. $D$ tends to a vanishing value as $t \to \infty$.

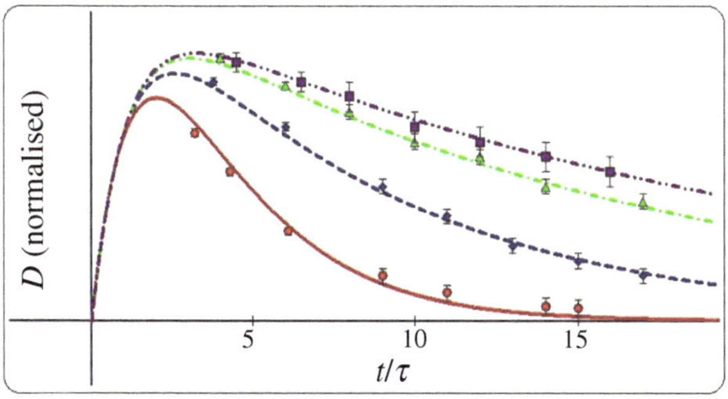


**Figure 13.8** Diffusion versus time (see text for details).

The results can be interpreted as follows: at early time, of the order of $\tau$, the system behaves as Drude-like, irrespective of $\omega_0$, with carriers assuming large mobility values. At increased times, charges become progressively localised as a result of scattering, significantly in agreement with the conclusions drawn by THz spectroscopy.

(i) *Quantum case*: It has been considered data of carbon nanotubes deriving by literature. We have three weights and search for the behaviour of $\langle \mathbf{v}(0) \cdot \mathbf{v}(t) \rangle$.

Parameters associated to the first weight $f_0$:

$f_0 = 0.064$; $\omega_0 = 0$; $\tau_0 = 7 \times 10^{-15}$ s.

Parameters associated to the second weight $f_1$:

$f_1 = 0.248$; $\omega_1 = 6.59 \times 10^{-12}$ Hz; $\tau_1 = 4.15 \times 10^{-14}$ s.

Parameters associated to the third weight $f_2$:

$f_2 = 0.176$; $\omega_2 = 1.166 \times 10^{-15}$ Hz; $\tau_2 = 3.7 \times 10^{-15}$ s.

The first weight has a contribution always positive and tending to zero at large times, the second weight produces a contribution initially positive and then becoming negative, tending to zero at very long times. The third weight gives an oscillating damped contribution. In Fig. 13.9, it is shown $\langle \mathbf{v}(0) \cdot \mathbf{v}(t) \rangle$ versus time considering the superposition of the three weights $f_0$, $f_1$ and $f_2$.

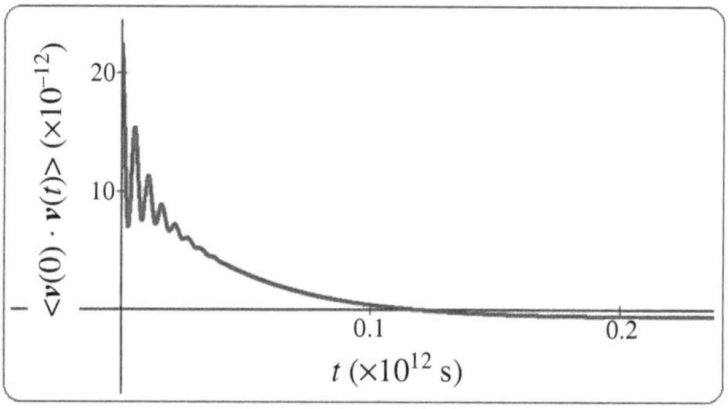


**Figure 13.9** Quantum case; superposition of states (see text for details).

Figure 13.10 shows the details at the beginning of the process.

(j) *Relativistic case*: it has considered the motion of electrons, at different velocities, inside a nanostructure of ZnO. In Fig. 13.11, we have $\langle \mathbf{v}(0) \cdot \mathbf{v}(t) \rangle$ versus $t$ for fixed value $\alpha_{R_{rel}} = 5$; the considered velocity of the carrier is $v = 10^7$ cm/s (solid line), $v = 10^{10}$ cm/s (dashed line) and $v = 2 \times 10^{10}$ cm/s (dots line), with $\tau = 0.84 \times 10^{-13}$ s and $T = 300$ K. The classical 'Drude' velocity $v = 10^7$ cm/s implies a negligible variation in mass for the electrons.

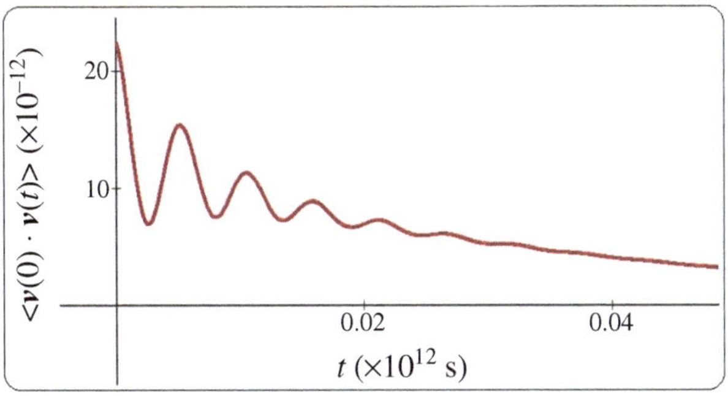


**Figure 13.10** Detail of beginning of process of Fig. 13.9.

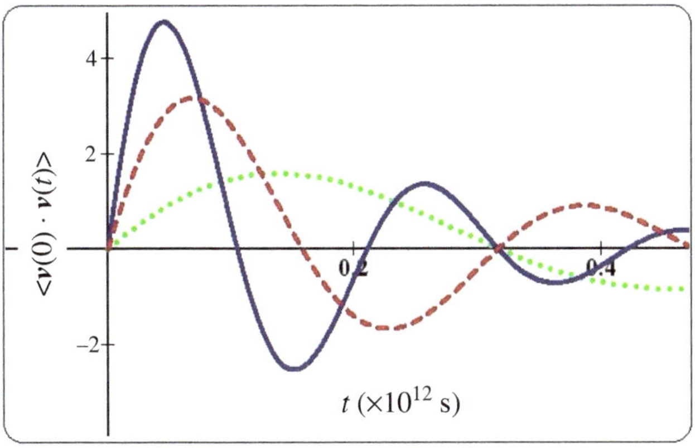


**Figure 13.11** Velocities correlation function versus time (relativistic case) (see text for details).

(k) *Relativistic case*: Figure 13.12 shows the behaviour of $D(t)$ versus $t$ for ZnO nanowires; $v_1 = 10^7$ cm/s, (dashed line), $v_2 = 10^{10}$ cm/s, (dots line), $v_3 = 1.5 \times 10^{10}$ cm/s, (dot-dashed line), $\omega = 10^{13}$ s$^{-1}$ and $T = 300$ K.

(l) Figure 13.13 represents the evolution of $R^2(t)$ versus time for electrons inside a ZnO nanostructure in relation to the classical 'Drude' velocity $v_1 = 10^7$ cm/s. $v_1$ implies a negligible variation of the electron mass. The parameters are $\omega_0 = 10^{13}$ s$^{-1}$

(fixed), $T$ = 300 K, $\tau_1$ = 0.28 × $10^{-13}$ s (ZnO nanowires, solid line), $\tau_2$ = 0.35 × $10^{-13}$ s (ZnO films, dashed line) and $\tau_3$ = 0.39 × $10^{-13}$ s (ZnO annealed films, dots line).

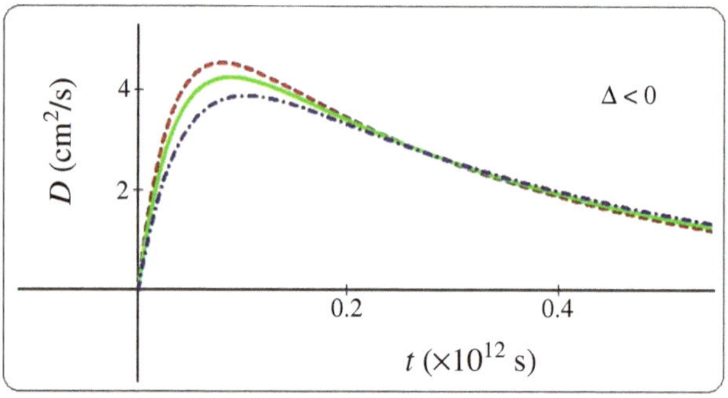


**Figure 13.12** Diffusion versus time (relativistic case) (see text for details).

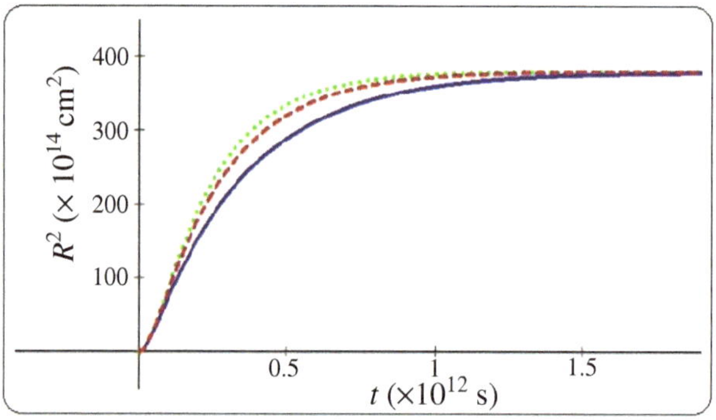


**Figure 13.13** $R^2$ versus time (relativistic case) (see text for details).

(m) Figure 13.14 represents the behaviour of diffusion versus time for CNTs at three different temperatures, with $m^*$ = 0.5 $m_e$, $\tau$ = 0.17 × $10^{-12}$ s and $\alpha_I$ = 0.5.

(n) Figure 13.15 shows the variation of the diffusion in time for CNTs, starting by non-relativistic velocity of electrons and then considering the relativistic velocities. It is: $v_e$ = $10^7$ cm/s (solid line), $v_e$ = $10^{10}$ cm/s (dots line), $v_e$ = 2 × $10^{10}$ cm/s (dot-dashed line), $\alpha_{I_{rel}} = 0.5$, $m^*$ = 0.5 $m_e$ and $\tau(s)$ = 0.17 × $10^{-12}$ s.

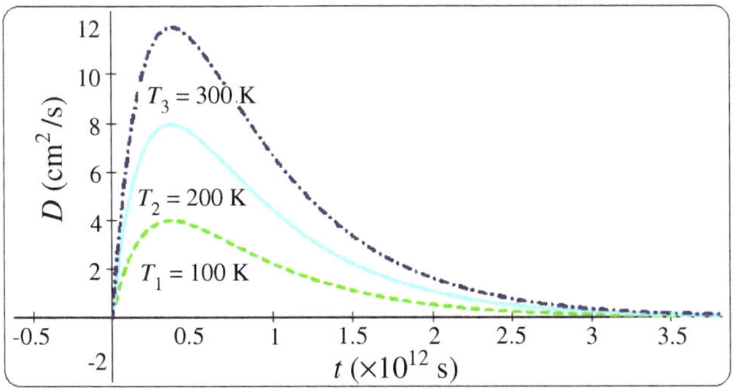


**Figure 13.14** Behaviour of diffusion versus time (see text for details).

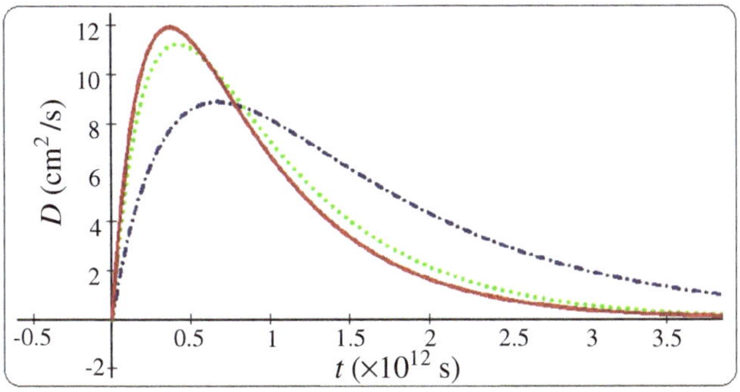


**Figure 13.15** Behaviour of diffusion versus time (see text for details).

(o) Figure 13.16 shows the behaviour of $<\boldsymbol{v}(t) \cdot \boldsymbol{v}(0)>_T$ versus $t$ for the indicated nanomaterials. The considered parameters are: $T = 310$ K, $\alpha_l = 0.5$, an average relaxation time $\tau_{\text{av}} = 10^{-13}$ s and the values of $m^*$ as resumed in Table 13.1.

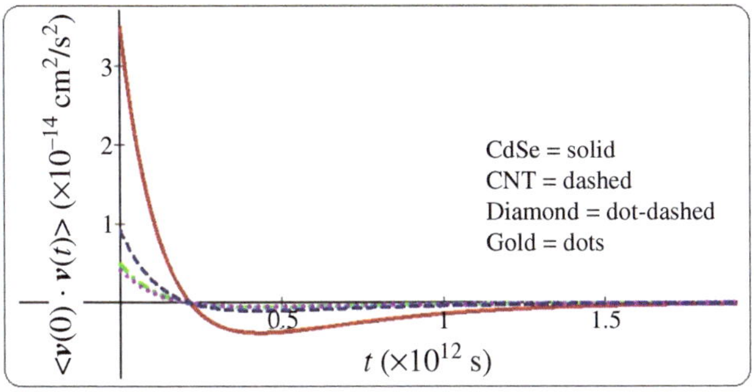


**Figure 13.16** Velocities correlation function versus time for different nanomaterials (see text for details).

**Table 13.1** Effective mass of the indicated nanomaterials

| **Material** | $m^*$ |
|---|---|
| CNTs | 0.5 $m_e$ |
| Gold | 1.1 $m_e$ |
| CdSe | 0.13 $m_e$ |
| Diamond | 0.94 $m_e$ |

# Chapter 14

# Plasmonics and Modelling

## 14.1 Introduction

Many nanostructures, in particular nanostructured metals, show very particular and interesting optical properties. One of the most promising phenomena encountered in these structures is the electromagnetic resonance, due to collective oscillations of the conduction electrons, said *plasmons*. Plasmon modes exist in different geometries and in varied metals, in particular in noble metals such as gold and silver. Under special circumstances, plasmons are excited by light; this leads to strong light scattering and absorption and increase of the local electromagnetic field. The interest in plasmon modes started at the beginning of the 20th century, with Zenneck (1907), Mie (1908), Sommerfeld (1909) and other scientists; recent advances regarding structure, manipulation and observation at the nanoscale increased the study of this field. Theoretical efforts encountered also the growing demand at a technological level, in particular for semiconductor based integrated electronic components, optical applications and new nano-components. This field brings important challenges for research and development processes, as the guide of light in integrated optical systems and the interface with electronic components. Many nanostructures are believed to be key ingredients for future optoelectronic devices.

*Mathematics and Physics for Nanotechnology: Technical Tools and Modelling*
Paolo Di Sia

ISBN 978-981-4800-02-0 (Hardcover), 978-0-429-02775-8 (eBook)
www.panstanford.com

## 14.2 Plasmons

The concept of *plasma* is very useful in the description of some aspects of the interaction radiation or conductive matter; free electrons of the conductive material (e.g., a metal) are considered as an electron fluid with high density, of the order of $10^{23}$ $cm^{-3}$. Such concept is the basis of the classic Drude model, which assumes that the matter contains stopped positive ions and a gas of classical not interacting electrons, whose motion results slowed by a 'viscous friction' force, due to the collisions with ions and characterised by a relaxation time $\tau$. The motion of electrons results casual, due to the continuous collisions with the lattice. The *plasma frequency* is defined as the proper frequency of the collective motion of electrons in the following way:

$$\omega_p^2 = \frac{\sigma_0}{\varepsilon_0 \tau} = \frac{ne^2}{\varepsilon_0 m}, \tag{14.1}$$

where $\sigma_0$ is the conductivity of material and it holds the common meaning of the other mathematical symbols. The dielectric constant of metal can be written as a function of the plasma frequency; in first approximation it results:

$$\varepsilon(\omega) = 1 - \frac{\omega_p^2}{\omega^2}. \tag{14.2}$$

By comparison between $\omega$ and $\omega_p$, we can deduce the behaviour of the electromagnetic waves coming to the metal.

If it holds $\omega < \omega_p$, the dielectric constant is negative, therefore, its square root is imaginary pure; this involves the *reflection* of the incident wave.

In the contrary case, the square root of the dielectric constant is real and the incident wave can *propagate* in the medium with a small attenuation.

The *plasma oscillations* are fluctuations of charge density, which happen to the frequency $\omega_p$ and propagate inside the metal. The quanta of such fluctuations inside the volume are said *volume plasmons*.

There is also a mechanism on the metal surface, characterised by the *surface plasmons*. Normally, it appears in the wavelength range between the visible and the infrared for the interface air or metal.

Localised surface electromagnetic excitations can exist at the surface of nanoparticles and metallic nanostructures: the *localised surface plasmons* (LSP). The frequency and the intensity of the radiation are very sensitive with respect to dimension, shape and morphology of the nanostructure. Nanoparticles and nanostructures with smaller dimensions with respect to the wavelength of the exciting light are characterised by a wide absorption band, normally in the range of the visible and near infrared spectrum.

## 14.3 Related Theoretical Models

Models concerning plasmonics simulate the extinction of the localised surface plasmons resonance (LSPR) by nanoparticles and nanoarrays; the application concerns the calculation of the light absorption and the scattering. About the related models, we have:

(a) The Mie theory
(b) The Gans theory
(c) The discrete-dipole approximation method (DDA)
(d) The finite-difference time-domain method (FDTD)
(e) The Di Sia model (DS model)

### 14.3.1 The Mie Theory

The Mie theory is a theoretical approach concerning the optical properties of nanoparticles. When the nanoparticle dimension is smaller than the wavelength of the incident light, such theory predicts that the extinction caused by a metallic nanosphere is estimated in the quasi-static and dipole limit. The relative cross-section of the process results:

$$\sigma_{\mathrm{ext}} = \frac{24\pi N_{\mathrm{A}} a^3 \varepsilon_{\mathrm{m}}^{3/2}}{\lambda \ln(10)} \left[ \frac{\varepsilon_{\mathrm{i}}}{(\varepsilon_{\mathrm{r}} + \chi \varepsilon_{\mathrm{m}})^2 + \varepsilon_{\mathrm{i}}^2} \right], \tag{14.3}$$

where $N_{\mathrm{A}}$ is the surface density, $a$ is the radius of the metallic nanosphere, $\varepsilon_{\mathrm{m}}$ is the dielectric constant of the medium surrounding the nanosphere, $\lambda$ is the wavelength, $\varepsilon_{\mathrm{i}}$ and $\varepsilon_{\mathrm{r}}$ are imaginary and real part of the dielectric function of the metallic nanosphere, respectively, and $\chi$ is a form factor concerning the nanoparticle. The localised resonance depends also on the interparticle space and on the dielectric constant of the substrate.

### 14.3.2 The Gans Theory

The Gans theory extends the Mie theory to the spheroidal particles case. In the case of polarisation of incident light parallel to the symmetry axis of the spheroid, the cross section of the process is given by:

$$\sigma_{\text{ext}} = \frac{2\pi V \varepsilon_{\text{m}}^{3/2}}{3\lambda} \sum_j \left[ \frac{(1/P_j)^2 \varepsilon_j}{(\varepsilon_{\text{r}} + \frac{1-P_j}{P_j}\varepsilon_{\text{m}})^2 + \varepsilon_j^2} \right], \tag{14.4}$$

where $V$ is the volume of nanoparticle and $P_j$ are the depolarisation factors along the three Cartesian axes, which hold into account of the anisotropic form of particles.

### 14.3.3 The Discrete-Dipole Approximation Method (DDA)

With this method, the nanoparticles are divided in a cubic array of $N$ polarizable dipoles, with polarizabilities $\alpha_i$ determined by the dielectric function of nanoparticles. The induced dipole $P_i$ of every element results $P_i = \alpha_i E_{\text{loc},i}$, in an applied plane wave field. The cross-section is determined by:

$$\sigma_{\text{ext}} = \frac{4\pi k}{|\boldsymbol{E}_{\text{inc}}|^2} \sum_{j=1}^{N} \text{Im}\left(\boldsymbol{E}_{\text{inc},j}{}^* \cdot \boldsymbol{P}_j\right), \tag{14.5}$$

with $k = \omega/c$ (in vacuum). Such method can be applied for the calculation of the absorption, scattering, extinction and other optical properties of nanoparticles with various forms and dimensions. It is considered one of the most important methods for the understanding of the structural characteristics and optical properties of nanomaterials and nanostructures.

### 14.3.4 The Finite-Difference Time-Domain Method (FDTD)

This is a method of numerical calculation in order to resolve the Maxwell equations directly in the time domain. Being a time-domain method, the solutions can cover a wide frequency range with a single process of simulation. It is a versatile and useful technique for applications where the resonance frequencies are not exactly known. A great variety of magnetic and dielectric materials can be modelled in a relatively simple way with such method.

### 14.3.5 The DS Model

The DS model, as a generalisation of the Drude formula based on linear response theory and resonant plasmonic mode, is able to accommodate many observed phenomena and to give a detailed description of the dynamic response of carriers. One of the principal features is the assumption of a characteristic frequency describing a collective mode and the evaluation of the response in the time domain by the residue theorem in the complex frequency plane. Experimental data have indicated that plasmon models describe nanostructured systems in a particularly effective way.

If we consider the conductivity, expressed as:

$$\sigma(\omega)=\frac{i\omega Ne^2}{m(\omega^2-\omega_0^2+i\omega/\tau)}, \tag{14.6}$$

in the terahertz (THz) range, it leads to a class of Drude–Lorentz functions.

The generalisation of Drude formula has interesting advantages, as the disposal of exact analytical results for describing the dynamic behaviour.

The results of the model can give an explanation of the ultra-short times and of high mobilities, with which charges spread in mesoporous systems that are of great interest in photocatalitic and photovoltaic systems. The relative short times, of the order of few relaxation time $\tau$, with which charges can reach much longer distance than typical dimensions of nanoparticles, indicate easy diffusion inside them. The unexplained fact found experimentally of ultra-short injection of charge carriers, in particular in Grätzel's cells, can be related to this phenomenon.

The Smith model is obtained as a limit of the DS model when $\alpha_I \rightarrow 0$. The situation $\alpha_I \rightarrow 0$ corresponds to $\tau \approx 1/2\omega_0$. From the other hand, both Smith's and DS models reduce to the Drude model in the limit $\omega_0 \rightarrow 0$. Therefore, although the two models are analytically different, their predictions are expected to be quite similar. The backscattering mechanism invoked by Smith arises in a natural way in the DS model, without further assumption on successive scattering events. The Smith model assumes that anisotropic scattering occurs only upon the first scattering event, while all subsequent events are characterised by isotropic scattering, without proposing physical reasons to justify this assumption.

The DS model deduced a ratio of two different relaxation times $\tau_1$ and $\tau_2$, which is $\tau_1/\tau_2 = (1 - \alpha_I) / (1 + \alpha_I)$.

For ZnO nanoparticles, investigated in THz experiments, double exponential decay indicates characteristic times $\tau_1 \approx 94$ ps and $\tau_2 \approx 2.4$ ns. This implies $\alpha_I = 0.92$, from which we deduce $\omega_0\tau \cong 0.2$. For $\tau \cong 10^{-13}$ s, we obtain $\omega_0 \cong 1.5 \times 10^{12}$ Hz, which is of the correct order of magnitude for the resonance in the infrared. The same procedure for injection times of films (500 nm grains) leads to $\omega_0 \cong 2.5 \times 10^{12}$ Hz (for $\tau \cong 10^{-13}$ s).

In GaAs nanowires, a resonance at $\omega_0 = 0.3$–$0.5$ THz explained the frequency dependent conductivity obtained by THz experiments. A long characteristic time $\tau_c \approx 1.1 \div 5 \cdot 10^{-12}$ s has been obtained, which according to DS model should correspond to $\tau_c = \tau/(1-\alpha_I) = 5 \times 10^{-12}$ s, leading to $\alpha_I = 0.1 \div 0.8$ for $\tau = 10^{-12}$ s and $\omega_0\tau = 0.3 \div 0.49$.

Similar considerations can be applied to single walled carbon nanotubes, where the resonance state is at $\omega_0 = 0.5$ THz with Drude–Lorentz scattering time $\tau \approx 10^{-13}$ s and to sensitised $TiO_2$ nanostructured films.

# Chapter 15

# Nanodiffusion in Graphene

## 15.1 Introduction

Currently, the global need of energy requires scientific and technological efforts for the development of nanodevices and nanosystems of critical importance in the sectors of implantable, ultrasensitive, bio-chemical, molecular, remote and mobile environmental sensoristics, medical science, nano-robotics, micro-(nano)-opto-electro-mechanical sector [M(N)OEMS], portable and resistant electronics, defence technology. The exploration of innovative nanotechnologies is therefore crucial for the conversion of mechanical, vibrational, hydraulic and chemical energies into electrical energy.

Nanotechnologies provide the control and manipulation of matter at the nanoscale through the use of properties and chemical-physical phenomena, manifested at this scale. A lot of empirical technologies, discovered in past years, have been understood and addressed to structures and mechanisms at nanometric scale.

Moreover, considering the global pollution, the research of green and renewable energy resources is one of the most urgent challenges to the sustainable development of human civilisation. At the macroscale, alternatives of the well-known energy resources such as oil, coal, hydroelectric energy, natural gas and nuclear energy are growing, for example, solar energy, geothermal energy,

*Mathematics and Physics for Nanotechnology: Technical Tools and Modelling*
Paolo Di Sia

ISBN 978-981-4800-02-0 (Hardcover), 978-0-429-02775-8 (eBook)
www.panstanford.com

biomasses, eolic energy and hydrogen. For all the indicated reasons, one of the challenges of current nanotechnology is the global study and development of ultrasmall, ultrasensitive, multifunctional, low-consumption systems.

Among the most commonly studied, promising and used materials at the nanoscale, we have zinc oxide (ZnO), titanium dioxide ($TiO_2$), gallium arsenide (GaAs), silicon (Si) and carbon nanotubes (CNTs). Considering the extraordinary thermal conductivity and electro-mechanical properties, carbon nanotubes have amazing utilisation for nanobiotechnology, electronics, optics and many other fields of materials science. The strength and flexibility of carbon nanotubes make them of potential use in controlling other structures at the nanoscale, particularly in nanobiotechnology engineering sector, as nanoelectronic devices, field emitters, gas sensors, high-strength composites and for hydrogen storage.

At atomic scale, physics of electronic transport processes in semiconductors radically changes; components and interconnections become smaller with respect to two characteristic lengths:

(a) *The electrons mean displacement*, in consideration to the fact that at such distances, by ordinarily dissipative ohmic-type, the transport regime tends to become ballistic.

(b) *The de Broglie wavelength of electrons*, in consideration to the appearance of quantum effects, as the discretisation of energy levels.

The bottom-up procedure utilises the relevant progress of quantum mechanics of many body systems and the development of supercomputers, allowing to simulate the dynamic evolution of electrons and atoms in wider systems, typically of dimension of clusters composing the nanostructured materials. Therefore, the welding between theoretical/simulation activity and experimentation on real systems, in the past separated by considerable differences of length and scales of time, shows all its importance. It is, therefore, crucial to have a deep theoretical understanding of the involved phenomena, through existing and new models.

The interest for carbon nanotubes is due to their remarkable structural, chemical and physical properties, focused to the production of advanced technological devices. The peculiar electronic properties (a metallic or semiconductor system depending by the geometrical characteristics), mechanical properties (high mechanical resistance both to traction and torsion) and thermal properties of nanotubes make them particularly useful for the realisation of molecular diodes, emitting field systems and scanning probe microscopes. The integration of carbon nanotubes in composite structures takes advantage of their particular properties for the creation of complex systems with multiple functional characteristics.

Conductive polymers (CPs) are organic materials with an intrinsic electrical conductivity at room temperature, due to the extended π-electron system which enlarges across the main polymer chain. The conduction process is related to chemical or electrochemical oxidation reactions (more rarely to reduction reactions), which affect this chain and introduce charge centres responsible for the conductivity of material.

The CPs attention as new plastic materials with high electrical conductivity originated by the surprising discovery that polyacetylene, exposed to iodine vapours, develops a very high electrical and physically well-characterised conductivity ($10^5$ S/cm), despite of its high chemical air instability. Polyaniline, polypyrrole, polythiophene and all their derivatives become today the most extensively investigated CPs, both for scientific-technological and commercial reasons.

The reduction in size of additive from micro to nanoscale involves many advantages in terms of composite performance, essentially resulting by the nanosize of inclusions and therefore by the resulting increase of the total surface area of material.

Reducing the size of additive to some orders of magnitude, it is possible to take advantage of new physical phenomena caused by induced quantum confinement effects. In scientific literature, we can get interesting discussions about possible physical manifestations related to the reduction in size of additives in composite systems. Therefore, the indisputable advantage of the use of nanometric systems as basic constituents for composite materials mainly improves the performance of a material with low loading levels of additive.

Comparing size, shape, properties and main applications of traditional and nanoscale additives, we note how very few traditional additives have dimensions of the order or less than

100 nm and how high shape factors values are essentially characteristic of nanostructured systems.

Since 2001, about research projects of nanocomposite materials, including metals, ceramics and thin films, polymer nanocomposites (PNCs) constitute more than 43% of the total, with attention to nanoclays and nanotubes carbon-based PNCs, representing almost 50% of the research. The interest in polymer-carbon nanotubes composites raised in relation to the nanotubes availability.

Nanostructured materials have contributed to a rapid PNCs development; further studies are necessary to fully understand the macroscopic properties of a given nanocomposite from its basic constituents.

It is desirable to develop new tools for machining and control, able:

(a) To mix and uniformly distribute individual nanostructures in the polymer matrix, in order to obtain homogeneous bulk properties;
(b) To induce a chemical compatibility among the nanostructures and the polymer medium;
(c) To vary the concentration of the inclusions inside the matrix, according to specific criteria.

Taking inspiration by biological systems, the idea is to prepare multifunctional structures acting as valid alternative to active components, usually employed in traditional devices as batteries, sensors, actuators and so on. The low cost additives might have particular advantages in the production of consumer goods, while the more expensive additives could have a broader market in the industry of high quality, as medical and electronics industries. Therefore, PNCs are becoming an integral part of modern technology devices; further studies are needed to understand how the enormous potential of this class of materials can be highlighted.

## 15.2 Peculiar Properties of Carbon Nanotubes

Carbon nanotubes are composed exclusively by carbon atoms, arranged in regular and symmetric crystals and grouped into the broader class of carbon nanostructures, which includes also the fullerenes, the onion-like, etc.

The structures formed by a single graphite rolled plane are known as *single-walled carbon nanotubes* (SWCNTs) or *graphene tubes*, while those formed by more rolling up tubes are called *multi-walled carbon nanotubes* (MWCNTs) (Figs. 15.1 and 15.2).

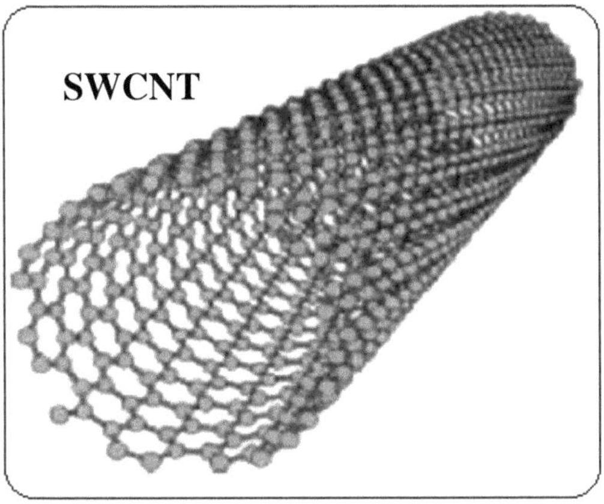


**Figure 15.1** Single-walled carbon nanotube.

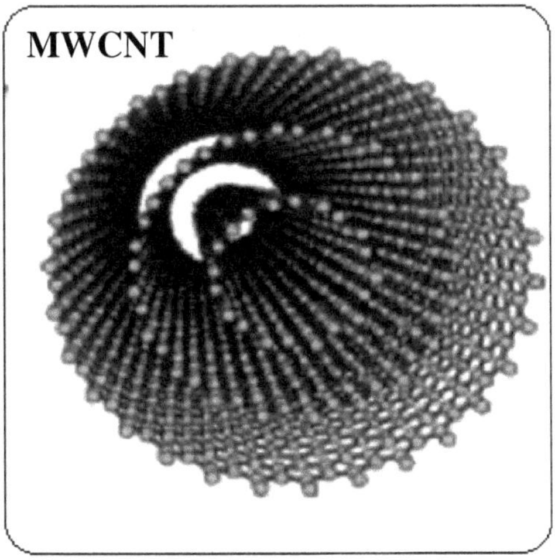


**Figure 15.2** Multi-walled carbon nanotube.

In MWCNTs, the radial distance between two concentric cylinders is typically of the order of 3.4 Å, the 3% greater than the interplanar spacing of graphite. The tubes diameter can vary by a few nanometres to a few tens nanometres, while the length can reach even values of tens of microns. The majority of nanotubes appear closed at the ends.

In the case of SWCNTs, the closure is realised by elements of fullerene type (half-spheres composed by pentagonal and hexagonal regularly arranged rings); for MWCNTs, the closure is geometrically more complex, even remaining highly symmetric and regular.

As a result of van der Waals forces, graphene tubes clump together to form beams in which the tubes are arranged in a triangular lattice with lattice constant of the order of 17–20 Å and intertubular distance of about 3 Å.

### 15.2.1 Structure

Considering the simple and well-defined structure, single-walled nanotubes are an ideal prototype of one-dimensional systems. The helicity is the measurement describing the possible arrangements of the hexagonal lattice of the plane with respect to the main axis of the tube; it is determined by the pair of integers $n$ and $m$ that identify the chiral vector $C_h = (n, m)$. By the magnitude of the chiral vector, the length $L$ of the circumference of a SWCNT can be obtained as:

$$L = a\sqrt{n^2 + m^2 + nm}\,, \tag{15.1}$$

where $a = 0.246$ nm is the lattice constant of graphite. By the length $L$ of the circumference, the tube diameter $D$ can be obtained through relation $D = L/\pi$.

The orientation is given by the chirality angle $\theta$:

$$\sin\theta = \frac{\sqrt{3}\,m}{2\sqrt{n^2 + m^2 + nm}}\,, \tag{15.2}$$

with $0^0 \leq \theta \leq 30^0$.

Depending on the values assuming by $n$, $m$ and $\theta$, we can distinguish three types of single-walled nanotube:

(a) $m = 0 \rightarrow$ 'zig zag' nanotube

(b) $m = n \rightarrow$ 'armchair' nanotube

(c) $(n, m)$ with $m \neq 0 \rightarrow$ 'chiral' nanotube

### 15.2.2 Synthesis

The most used synthesis techniques for the preparation of nanocrystalline carbon structures can be grouped into four main classes:

(a) Laser ablation
(b) Arc discharge
(c) Thermal pyrolysis
(d) Chemical vapour deposition (CVD)

Carbonaceous precursors are used, both in solid and in gaseous phase and the reactions which take place during the synthesis generally occur at high temperatures. Very different structures can be produced, depending on values of temperature and pressure, density of the precursors and of the chemicals around, which the aggregation process occurs. The presence of other species can significantly affect the growth and determine the type of final structures. Transition metals such as iron, cobalt, nickel and yttrium favour the nucleation and growth of SWCNTs and are normally used to increase their efficiency.

### 15.2.3 Electronic Properties

By many points of view, single-walled nanotubes can be considered a nanoscale version of carbon fibres, showing unique physical and chemical characteristics. The electronic properties of nanotubes derive directly from those of the graphenic plane when this is rolled to form the tubular structure. Electrons are constrained to move in a periodic potential and may assume only a small subset of wavelengths, or quantum states, which depend on the diameter and chirality of the formed tube, as result of rolling.

Along the direction of chiral vector (along the circumference), energy levels are discrete, while along the nanotube axis the density of states tends to continuous and electrons are therefore moving freely. So, despite the structural similarity to graphite plane, which is properly a 'half-metallic' material, SWNTs can assume metallic or semiconducting behaviour, depending on the mode in which the plan is rolled up to form the tube. In nanotubes without defects, the electrons can ballistically travel inside the tube, i.e., in absence of ordinary scattering phenomenon, responsible for the electrical resistance of material.

Thanks to these properties, nanotubes are ideal systems for the production of quantum wires, leading to the change by microelectronics to nanoelectronics and overcoming the barriers of miniaturisation and heat dissipation related to the silicon technology. However, only a part of nanotubes has adequate diameter and chirality for reaching the Fermi level in the subset of allowed states and therefore for showing metallic-type behaviour. The conduction properties of nanotubes can be varied even with doping of material, by inserting atoms as nitrogen and boron into the tubes structure.

The good electrical properties and the elongated shape of nanotubes make them an almost ideal class of electron emitters. Given their pointed thread-like geometry, nanotubes emit electrons at low potential, if compared to conventional conductive materials; moreover the strong bonds among carbon atoms allow the emission process for long time, meanwhile they maintain the structural integrity of the system. The applications range spans from the realisation of flat screens to electronic components, switches, fluorescent lamps, electron microscopes and so on.

Carbon nanotubes have shown to be extremely sensitive to intense electric fields; they react by bending up to 90° and regain the original shape as soon as the electric field is interrupted, without damages. If we apply an oscillating electric field, nanotubes vibrate and carefully controlling the oscillation frequency, it is possible to put them in resonance. Experiments in this direction showed that each nanotube has a precise resonance frequency, depending on length, diameter and morphology. This interesting property gave rise to many applications, such as the creation of nanobalances, measuring weight of bodies in the order of femtograms.

Another interesting property of nanotubes is the modulation of electric resistivity as a function of mechanical deformation, commonly known as *piezoresistive effect*; it has been found both in the single nanotube and on SWCNT-based macroscopic systems.

SWCTs-made membranes provided gauge factor values 2.2 times higher than that of silicon, usually utilised for the production of piezoresistive devices. This property can effectively help in the realisation of lightweight and highly sensitive pressure sensors.

### 15.2.4 Mechanical Properties

Nanotubes have high flexibility and rigidity; experimental evidence shows that they may give rise to rigid elastic systems with very low density and interesting mechanical properties of tensile strength and Young's modulus. Theoretical studies indicated extremely high values of Young's modulus, up to 5 TPa.

Another very fascinating property is the ability of nanotubes to withstand high levels of plastic deformation without fracturing, combined with their low density (1.33–1.40 g/cm$^3$); it is a big point of advantage if compared to available carbon fibres.

Nanotubes can also undergo up to 16% axial deformations without fractures; simulation studies expected the possibility to induce reversible deformations as bending, torsion and axial compression up to 30–40% before fractures. The insertion of nanotubes in a polymer material may lead to an increase of the system resistance; this result depends on several factors, such as the ability to uniformly disperse the nanotubes in the matrix and generate a good nanotube-matrix adhesion, so as to effectively transmit the effort and to avoid the sliding of SWCNT bundles.

### 15.2.5 Thermal Properties

The value of thermal conductivity at room temperature for a SWCT, estimated by molecular dynamics calculation, is of the order of $6.6 \times 10^3$ W/mK, greater than that of isotropically pure diamond ($3.32 \times 10^3$ W/mK), which is a material with exceptional thermal conductivity. This means that, at theoretical level, SWCNTs could be defined as the best heat conductor existing in nature.

Unlike metals, for which the thermal energy is mainly transmitted by electrons, in nanotubes the phonon component is dominant in heat conduction and consequently the thermal conductivity is proportional to the product of the specific heat times the sound speed times the phonon mean free path. Considering that SWCNTs have a long phonon mean free path (due to the absence of defects on atomic scale), a high transmission speed of sound ($\sim 10^4$ m/s) and a high specific heat (equal to or greater than that of diamond), high conductivity values are found.

Performed measurements with different experimental methods gave values of thermal conductivity decreasing when nanotubes are aggregated into bundles, possibly due to the decrease of the phonon mean free path. Nanotubes can therefore be successfully integrated in microelectronic devices, where would go to take advantage of the anisotropic properties of conduction and rapid dispersion of tubes heat for increasing the operating time of equipment.

### 15.2.6 Chemical and Electrochemical Properties

The SWCNTs have a high chemical inertia, are resistant to oxidising agents and acids and because of their tubular form, show strong properties of capillarity, which together with the high surface/weight ratio, make them ideal materials for the adsorption of liquids and gases.

The adsorption properties of nanotubes have been studied mainly in the case of hydrogen, in view of its interesting use in fuel cells, since all systems used for storage (cylinders, hydrides and activated carbons) require to work at high pressure and low temperature in order to store an adequate gas amount.

Nanotubes find applications also as electrodes of supercapacitors (electrochemical capacitors), i.e., batteries with the capacity of storing large energy amounts and characterised by very fast processes of charge and discharge. These properties are due to the high porosity and therefore to the high effective surface area characterising the electrodes of these devices. Nanotubes show to be ideal candidates for this kind of application, in view of their high surface area, estimated at $\sim 1.3 \times 10^3$ m$^2$/g.

### 15.2.7 Nanobiosensing Properties

Considering that the conductivity of nanotubes strongly depends by their structure, doping and environmental conditions, it is possible to use them also as chemical sensors. Nanotubes-based sensors can answer to absorption of particular gaseous species in terms of changes in electrical conductivity. The advantages arising by the use of nanotubes reside in the possibility of obtaining selective and reversible systems, able to operate at room temperature with very fast response times. Studies on sensors consisting of individual semiconducting nanotubes showed how these devices are responding in a very selective and rapid way to absorption of species such as $NO_2$ and $NH_3$.

Although the mechanisms of interaction among gaseous species and nanotubes require still further development, these devices represent a significant improvement in applicative fields; the conventional solid-state sensors for the determination of $NO_2$ and $NH_3$ generally operate at temperatures of the order of 400 °C, with a sensitivity $\sim 10^3$ times less than the new nanotube devices. One of the current problems lies in the great difficulty in building single-nanotube devices. Systems using aggregates of nanotubes have been created, showing how the sensor response is strictly dependent by different factors, such as the synthesis and the organisation of the deposit of nanotubes, their degree of purity, the temperature of the substrate supporting them and the possible device polarisation.

Assembled sensors and biosensors based on SWCNTs have been successfully used for the electrochemical determination of many chemical species of biological and environmental interest, as serotonin, NADH, catechol, glucose, epinephrine, ascorbic acid and uric acid. These sensors have an electrochemical activity and higher sensitivity if compared to conventional graphite electrodes and are also miniaturised and chemically inert.

## 15.3 Fields of Utilisation

Carbon nanotubes are $10^2$ times stronger than steel, among the best field emission emitters, can maintain a current density of the order or more than $10^{-9}$ A/cm$^2$ and have a thermal conductivity comparable to that of diamond. These characteristics make them potentially useful in a wide range of applications, from optics to nanoelectronics, from composite materials to conductive polymers, in sensoristics. The most basic building block of these systems is the electric wire, an excellent conductor. CNTs carry high currents with negligible heating, act as sensing materials in pressure, flow, thermal, gas, optical, mass, position, stress, strain, chemical and biological sensors.

Important sectors of CNTs utilisation are as follows:

(a) **Biomedical industry:** The glucose sensing applications and the monitoring of the exposure to hazardous radiation like in nuclear plants/reactors or in chemical laboratories or industries, as implantable sensors, used for tracking glucose levels in blood and for detecting DNA sequences in human body.

(b) **Food industry:** Sensors and biosensors for providing safety and quality control, for meat freshness monitoring. CNT-based chemical sensors can be used also for detecting undesired chemical residues resulting from food additives, animal drugs, pesticides, herbicides and other environmental contaminants.

(c) **Agriculture and fishing industry:** MWCNT-coated quartz crystal microbalance humidity sensors are used for monitoring relative humidity, CNT-based $CO_2$ sensors to monitor the $CO_2$ concentration, CNT-based pressure sensors for uniform spraying of liquid fertiliser, insecticides, pesticides and herbicides. CNT-based pH sensors are highly useful for maintaining proper pH balance of water quality for fishing industries.

(d) **Manufacturing industry:** CNT-based strain and pressure sensors for strain and pressure measurements in manufacturing industries, CNT-based hydrogen gas sensors in hydrogen monitoring and control for petroleum transformation, rocket engines and welding.

(e) **Security:** CNT-based sensors in defence and security, in unmanned defence systems like unmanned aerial vehicles, CNT-based chemical sensors as electronic noses for detecting hidden bombs, chemical weapons in luggage, vehicles and aircraft.

The exceptional properties of these compounds make them relevant to a very growing quantity of other potential interesting and useful applications.

## 15.4 Nanodiffusion: Classical and Quantum Results

We used the Di Sia model (DS model) for a variety of applications related to graphene. It follows some examples considered in detail.

### 15.4.1 Classical Case: Dependence on Temperature

We consider the behaviour of diffusion for CNTs at three different temperatures $T_1 = 100$ K, $T_2 = 200$ K, $T_3 = 300$ K, with $m^* = 0.5\ m_e$, $\tau = 0.17 \times 10^{-12}$ s and $\alpha_l = 0.5$. The diffusion coefficient $D$ is directly proportional to the temperature $T$, therefore, the increase of $T$ promotes the diffusion. With the chosen parameters, the values of $D$ span the range from 4–12 $cm^2/s$, of considerable technological and industrial sector interest (Fig. 15.3).

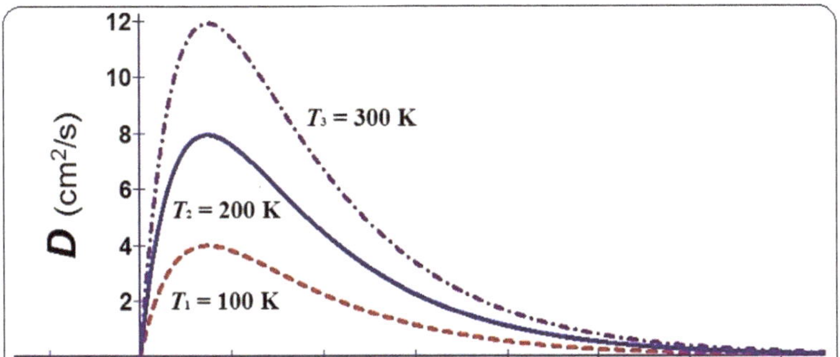


**Figure 15.3** *D* versus *t* for the indicated temperatures.

Comparing the values of diffusion of CNTs (blue dots line) (Fig. 15.4) with Si (red dashed line), $TiO_2$ (black dashed line), ZnO (light blue dots-dashed line) and GaAs (violet dashed line), it is possible to get interesting considerations. The used effective masses and relaxation times are summarised in Table 15.1.

**Table 15.1** Effective masses and relaxation times

| Nanomaterial | $m^*$ | $\tau$ (s) |
|---|---|---|
| CTN | $0.5\ m_e$ | $0.17 \times 10^{-12}$ |
| Si | $1.08\ m_e$ | $0.5 \times 10^{-13}$ |
| $TiO_2$ | $6\ m_e$ | $0.1 \times 10^{-13}$ |
| ZnO | $0.24\ m_e$ | $0.84 \times 10^{-13}$ |
| GaAs | $0.067\ m_e$ | $10^{-12}$ |

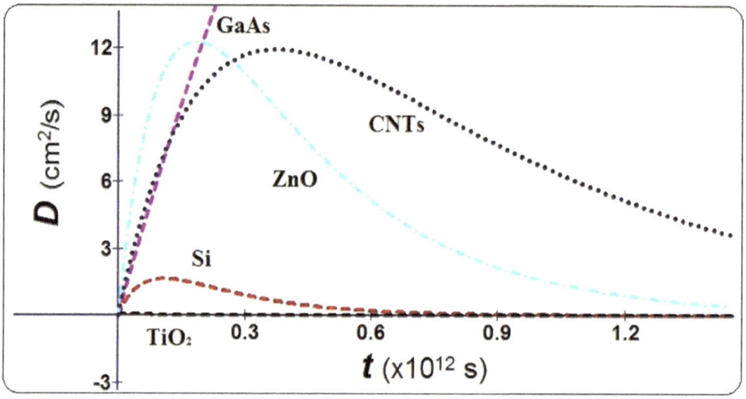


**Figure 15.4** *D* versus *t* for five important nanomaterials.

With the exception of GaAs (deliberately left out of scale in Fig. 15.4), CTNs present for short times a high diffusivity with respect to the other analysed nanomaterials, feature of great importance at nanobiosensoristic and nanotechnological level.

Figure 15.5 shows a zoom of Fig. 15.4 for *t*-axis, i.e., considering long times; only GaAs demonstrates to have a better response in diffusion than CNT, but with a peak at bigger time. This could be not interesting if we need a nanodevice with high fast initial diffusivity.

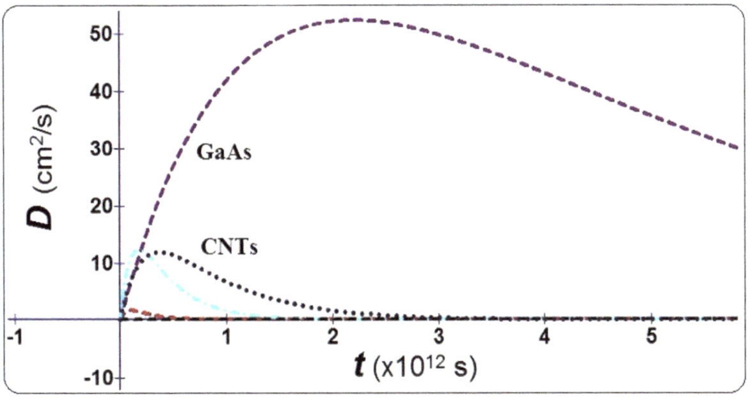


**Figure 15.5** Zoom of Figure 15.4 for *t*-axis considering long times.

## 15.4.2 Classical Case: Dependence on Chiral Vector

Considering changes of the effective mass of CNTs, it implies the variation of the chiral vector $C_h = (n, m)$. We have considered a great difference in mass and an intermediate value (Fig. 15.6), i.e., CNTs with $(n, m)$ indices equal to (3, 1) (red dots-dashed line), (4, 2) (light blue dashed line) and (9, 2) (light red dots line).

The decrease of the effective mass involves a raise of the diffusion in CNTs; the particular needs at practical and technological level will imply the choice of the CNT with the right indices.

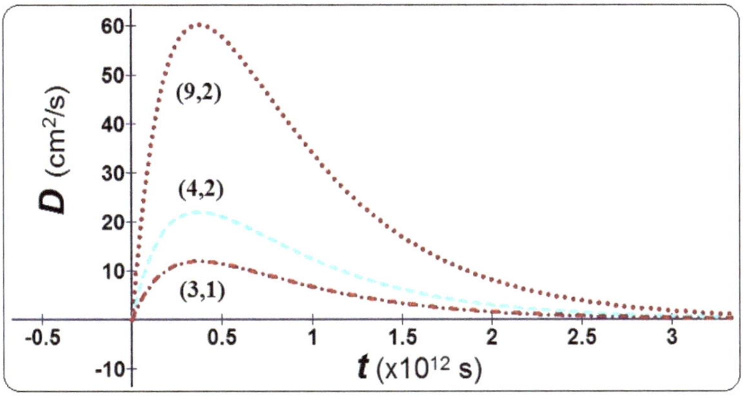


**Figure 15.6** *D* versus *t* for CNTs with three different chiral vectors.

### 15.4.3 Classical Case: Dependence on Relaxation Times

The variation of relaxation time affects the diffusion of CNTs. Considering the values $\tau_1 = 10^{-13}$ s (red dots line) (Fig. 15.7), $\tau_2 = 0.17 \times 10^{-12}$ s (blue dots-dashed line) and $\tau_3 = 10^{-12}$ s (light blue solid line), we note how, changing $\tau$, not only the value of $D$ changes, but also the width of the curves, since the change involves the arguments of the exponentials of Eqs. 13.64 and 13.68.

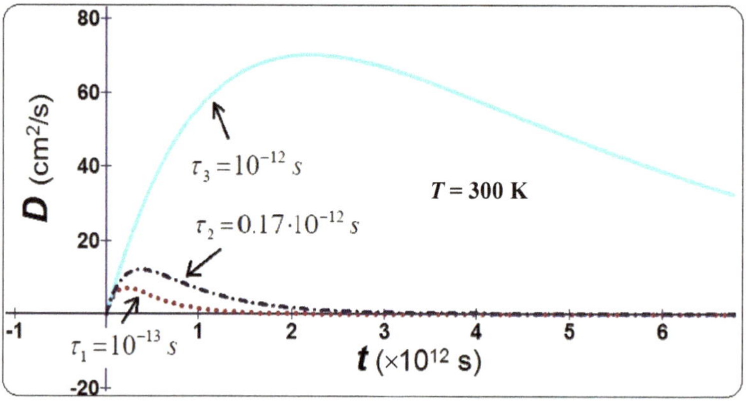


**Figure 15.7** $D$ versus $t$ for CNTs with three different values of $\tau$.

### 15.4.4 Quantum Case: Dependence on Temperature

At quantum level, with experimental data, it is possible to consider the desired (or needed) number of weights $f_i$ (Eq. 13.60) to keep into account for the implementation in Eqs. 13.75 and 13.79. It has been tested charge dynamics for single-walled carbon nanotube in transparent films, considering data, which have been elaborated and resumed in Tables 15.2 and 15.3.

We have three weights, bringing to a contribution in diffusion; the sum of these three parts gives the global behaviour of $D$.

Figures 15.8, 15.9 and 15.10 illustrate the behaviour of the diffusion coefficient $D$ versus time $t$, varying the temperature. We note also the presence of an initial damped oscillation, which is visible zooming the beginning of the process (Fig. 15.11). It is always present, irrespective of temperature and increases with temperature.

**Table 15.2** Data elaboration for the unbaked sample

| $T(K)$ | $\omega_{p_0}$ | $\tau_0$ | $\omega_{p_1}$ | $\omega_1$ | $\tau_1$ | $\omega_{p_2}$ | $\omega_2$ | $\tau_1$ |
|---|---|---|---|---|---|---|---|---|
| 100 | 69.42 | 0.010 | 132.24 | 0 | 0.249 | 68.88 | 187.35 | 0.020 |
| 200 | 69.54 | 0.011 | 131.43 | 0 | 0.240 | 69.54 | 187.23 | 0.019 |
| 300 | 69.06 | 0.013 | 130.11 | 0 | 0.231 | 69.84 | 187.32 | 0.020 |

*Note*: For the right interpretation of the table, it holds $\omega_{i\text{value}} = \omega_{i\text{table}} \times 10^{12}\,\text{Hz}$, $\omega_{p_i\,\text{value}} = \omega_{p_i\,\text{table}} \times 10^{12}\,\text{Hz}$, $\tau_{i\text{value}} = \tau_{i\,\text{table}} \times 10^{-12}\,\text{s}$. For peak zero, the centre frequency is zero.

**Table 15.3** Data elaboration for the baked sample

| $T(K)$ | $\omega_{p_0}$ | $\tau_0$ | $\omega_{p_1}$ | $\omega_1$ | $\tau_1$ | $\omega_{p_2}$ | $\omega_2$ | $\tau_1$ |
|---|---|---|---|---|---|---|---|---|
| 100 | 59.52 | 0.016 | 87.12 | 0.6 | 0.273 | 121.32 | 178.11 | 0.028 |
| 200 | 66.60 | 0.014 | 87.51 | 0.6 | 0.254 | 121.11 | 177.57 | 0.029 |
| 300 | 76.83 | 0.014 | 87.51 | 0.6 | 0.238 | 126.18 | 178.05 | 0.027 |

*Note*: For the right interpretation of the table, it holds: $\omega_{i\text{value}} = \omega_{i\text{table}} \times 10^{12}\,\text{Hz}$, $\omega_{p_i\,\text{value}} = \omega_{p_i\,\text{table}} \times 10^{12}\,\text{Hz}$, $\tau_{i\text{value}} = \tau_{i\,\text{table}} \times 10^{-12}\,\text{s}$. For peak zero, the centre frequency is zero.

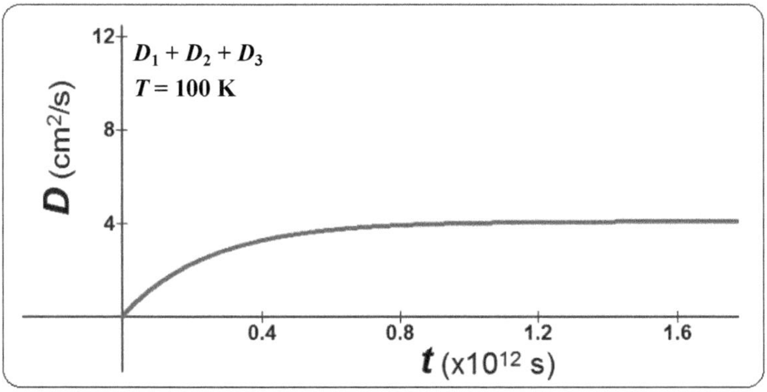


**Figure 15.8** Global diffusion versus time ($T$ = 100 K).

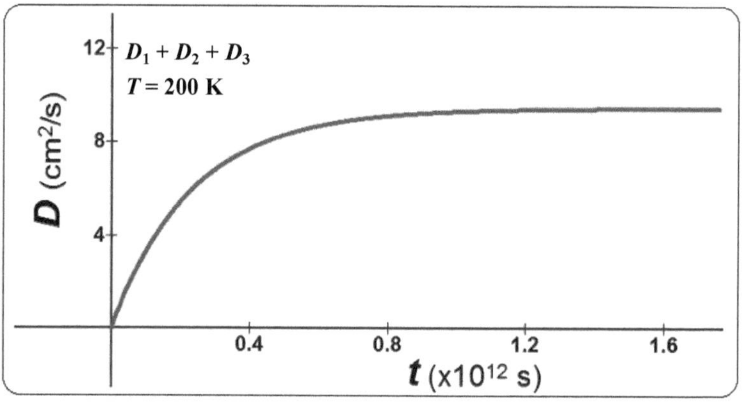


**Figure 15.9** Global diffusion versus time ($T$ = 200 K).

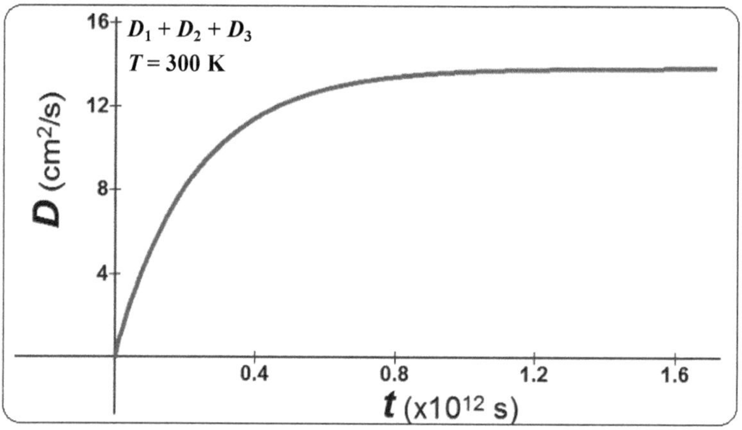


**Figure 15.10** Global diffusion versus time ($T$ = 300 K).

### 15.4.5 Quantum Case: Dependence on Baking

It has been carried out a thermal baking process for CNTs, treating samples in flowing pure argon at 1000 °C for 12 hours, by removing the nitric acid, used during its purification.

Figures 15.12, 15.13 and 15.14 illustrate the behaviour of the diffusion coefficient $D$ versus time $t$, varying the temperature.

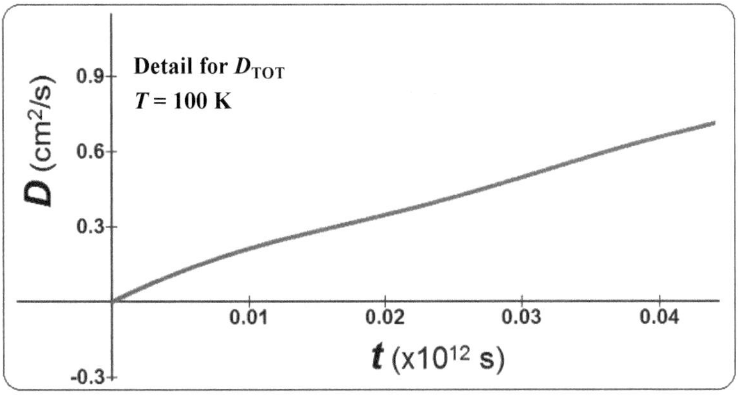


**Figure 15.11** Detail at beginning of the process for $D$ versus $t$ ($T$ = 100 K).

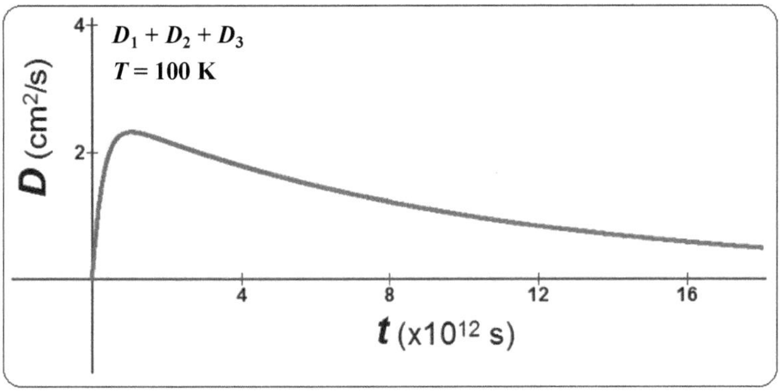


**Figure 15.12** $D$ versus $t$ for the baked sample ($T$ = 100 K).

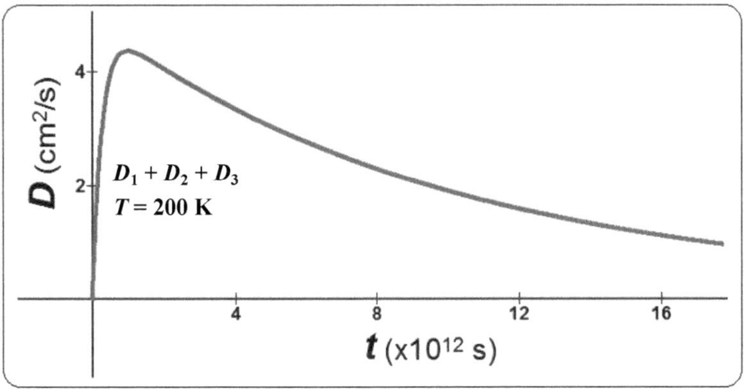


**Figure 15.13** $D$ versus $t$ for the baked sample ($T$ = 200 K).

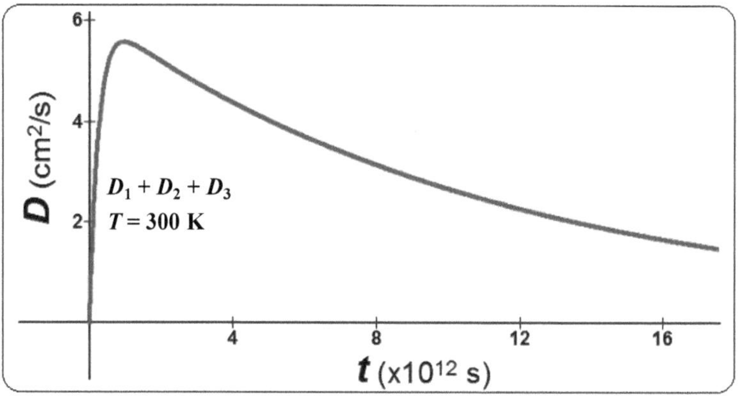


**Figure 15.14** *D* versus *t* for the baked sample (*T* = 300 K).

Also in this case, it appears an initial damped oscillation, which is visible zooming the beginning of the process (Fig. 15.15).

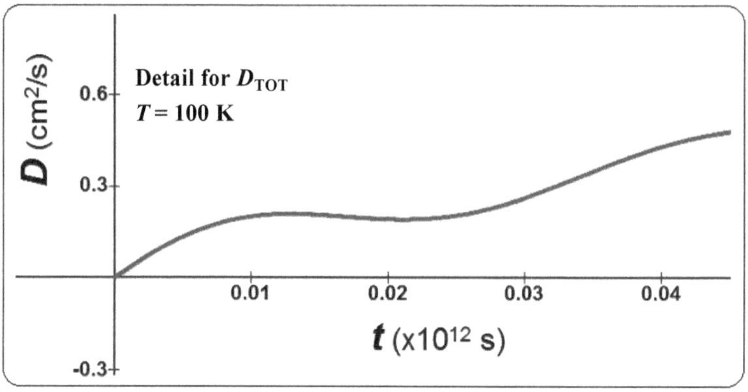


**Figure 15.15** Detail at beginning of the process for *D* versus *t* (*T* = 100 K).

## 15.5 Concluding Remarks

Tests of CNTs are showing that they represent a versatile and interesting nanomaterial, with high values of diffusion. They can be useful for a large spectrum of technological needs, in particular for nano-bio-devices and nanobiosensoristics, considering tailored combinations of the considered parameters.

The possibility to have a rapid answer for carriers transport is an important peculiarity for nano-energy systems and globally nano-technology area.

We underline:

(a) The appearance of oscillating damped contributions, that can have interesting implications for possible occurrence of drastic change of temperature for carbon nanostructure-made materials (e.g., for aeronautics and astronautics).

(b) The presence of small initial oscillations and marked initial peaks in diffusion can be an important support for nanobio-sensoristic applications, for monitoring, in nano-medicine, in situations in which a high sensitivity is required.

(c) The increase of temperature brings to a raise of the initial velocity and the initial peak of diffusion, allowing the design or creation of nanobiosensors, which work at room temperature.

(d) The oscillations of diffusion in time open the way to a search suggest at experimental level, for confirming these previsions. These initial features can also be considered as a further power test of current experimental time-resolved techniques, like time-resolved THz spectroscopy (TRTS).

# Further Readings


Altan, H., Huang, F., Federici, J. F., Lan, A., Grebel, H., Optical and electronic characteristics of single walled carbon nanotubes and silicon nanoclusters by tetrahertz spectroscopy, *J. Appl. Phys.*, 96, 6685 (2004).

Baxter, J. B., Schmuttenmaer, C. A., Conductivity of ZnO nanowires, nanoparticles, and thin films using time-resolved terahertz spectroscopy, *J. Phys. Chem. B*, 110, 25229 (2006).

Beard, M. C., Turner, G. M., Schmuttenmaer, C. A., Transient photoconductivity in GaAs as measured by time resolved terahertz spectroscopy, *Phys. Rev. B*, 62, 15764 (2000).

Bhushan, B. (Ed.), *Springer Handbook of Nanotechnology*, 3rd ed., Springer, Heidelberg (2010).

Borondics, F., Kamarás, K., Nikolou, M., Tanner, D. B., Chen, Z. H., Rinzler, A. G., Charge dynamics in transparent single-walled carbon nanotube films from optical transmission measurements, *Phys. Rev. B*, 74, 045431 (2006).

Bose, S., Quantum communication through an unmodulated spin chain, *Phys. Rev. Lett.*, 91, 207901 (2003).

Di Sia, P., *Appunti di Analisi Matematica 2*, CUSL - Il Sentiero, Verona (2005).

Di Sia, P., An analytical transport model for nanomaterials, *J. Comput. Theor. Nanosci.*, 8, 84 (2011).

Di Sia, P., An analytical transport model for nanomaterials: the quantum version, *J. Comput. Theor. Nanosci.*, 9, 31 (2012).

Di Sia, P., Nanotechnology between classical and quantum scale: applications of a new interesting analytical model, *Adv. Sci. Lett.* 5, 1 (2012).

Di Sia, P., A new theoretical model for the dynamical analysis of nano-bio-structures, *Adv. Nano Res.* 1, 1 (2013).

Di Sia, P., Looking at the Schrödinger equation for nanotechnology, *IJESIT*, 2(6), 410 (2013).

Di Sia, P., Mechanical behaviour at nanolevel: general aspects and modelling, *Int. J. Eng. Sci. Innov. Technol. (IJESIT)*, 2(6), 129 (2013).

Di Sia, P., About the influence of temperature in single-walled carbon nanotubes: details from a new Drude-Lorentz-like Model, *Appl. Surf. Sci.*, 275, 384 (2013).

Di Sia, P., *Fondamenti di Matematica e Didattica I*, Aracne Editrice, Roma (2013).

Di Sia, P., *Fondamenti di Matematica e Didattica II*, Aracne Editrice, Roma (2014).

Di Sia, P., Characteristics in diffusion for high-efficiency photovoltaics nanomaterials: an interesting analysis, *J. Green Sci. Technol.*, 1, 1 (2014).

Di Sia, P., Overview of Drude-Lorentz type models and their applications, *Nanoscale Sys. Math. Model. Theory Appl.*, 3, 1 (2014).

Di Sia, P., Present and future of nanotechnologies: peculiarities, Phenomenology, Theoretical Modelling, Perspectives, *Rev. Theor. Sci.*, 2(2), 146 (2014).

Di Sia, P., Relativistic nano-transport and artificial neural networks: details by a new analytical model, *IJAIM*, 3(3), 96 (2014).

Di Sia, P., Schrödinger equation and symmetry: an interesting overview, *J. Adv. Math. Appl. (JAMA)*, 3, 1 (2015).

Di Sia, P., A new analytical transport model for (nano)physics, *Int. Res. J. Eng. Tech.* (IRJET), 2(7), 1 (2015).

Di Sia, P., Quantum-relativistic carrier nano-transport and plasmonics, *FNP*, 1(1), 67–69 (2017).

Di Sia, P., Quantum-relativistic velocities in nano-transport, *Appl. Surf. Sci.*, 446, 187–190 (2018) (ISSN: 0169-4332), https://doi.org/10.1016/j.apsusc.2018.01.273

Dresselhaus, G. D. M. S., Avouris, P., *Carbon Nanotubes: Synthesis, Structure, Properties, and Applications*, Springer, Berlin (2002).

Gao, P. G., Song, J. H., Liu, J., Wang, Z. L., Nanowire piezoelectric nanogenerators on plastic substrates as flexible power sources for nanodevices, *Adv. Mater.* 19, 67 (2007).

Geim, A. K., MacDonald, A. H., Graphene: Exploring carbon flatland, *Phys. Today*, 60(8), 35 (2007).

Grady, B. P., *Carbon Nanotube-Polymer Composites: Manufacture, Properties, and Applications*, John Wiley & Sons Inc., New York (2011).

Graetzel, M., Photoelectrochemical cells, *Nature*, 414, 338 (2001).

Gratzel, M., Recent advances in sensitized mesoscopic solar cells, *Acc. Chem. Res.,* 42, 1788 (2009).

Hendry, E., Wang, F., Shan, J., Heinz, T. F., Bonn, M., Electron transport in $TiO_2$ probed by THz time-domain spectroscopy. *Phys. Rev. B* 69, 081101(R) (2004).

Kittel, C., *Introduction to Solid State Physics*, John Wiley & Sons, New York (2005).

Kroto, H. W., Fischer, J. E., Cox D. E. (eds.), *The Fullerenes*, Pergamon Press, Oxford (1993).

Liu, Z., Tabakman, S., Welsher, K., Dai, H., Carbon nanotubes in biology and medicine: In vitro and in vivo detection, imaging and drug delivery, *Nano Res.*, 2(2), 85 (2009).

Marulanda, J. M., Srivastava, A., Carrier density and effective mass calculation for carbon nanotubes, *Phys. Stat. Sol. (b)*, 245(11), 2558 (2008).

Nienhuys, H. K., Sundstrom, V., Influence of plasmons on terahertz conductivity measurements, *Appl. Phys. Lett.*, 87, 012101 (2005).

Parkinson, P., Joyce, H. J., Gao, Q., Tan, H. H., Zhang, X., Zou, J., *et al.*, Carrier lifetime and mobility enhancement in nearly defect-free core–shell nanowires measured using time-resolved terahertz spectroscopy, *Nano Lett.*, 9(9), 3349 (2009).

Parkinson, P., Lloyd-Hughes, J., Gao, Q., Tan, H. H., Jagadish, C., Johnston, M. B., *et al.*, Transient terahertz conductivity of GaAs nanowires, *Nano Lett.*, 7(7), 2162 (2007).

Patolsky F., Lieber, C. M., Nanowire nanosensors, *Mater. Today*, 8, 20 (2005).

Pirozhenko, I., Lambrecht, A., Influence of slab thickness on the Casimir force, *Phys. Rev. A*, 77, 013811–013818 (2008).

Reich, S., Thomsen, C., Maultzsch, J., *Carbon Nanotubes*, Weinheim: WILEY-VCH Verlag GmbH & Co. KGaA, New York (2004).

Rudin, W., *Real and Complex Analysis*, McGraw-Hill International Editions: Mathematics Series, McGraw-Hill Publishing Co, New York (1987).

Saito, R., Fujita, M., Dresselhaus, G., Dresselhaus, M. S., Electronic structure of chiral graphene tubules, *Appl. Phys. Lett.*, 60(18), 2204 (1998).

Sanguansri, P., Augustin, M. A., Nanoscale materials development—a food industry perspective, *Trends Food Sci. Technol.*, 17(10), 547 (2006).

Vlasov A. Yu., Information Nano-Technologies: Transition from Classical to Quantum, arXiv:0912.0829v1 [quant-ph] (2009).

Wang, Z. L., Nanostructures of zinc oxide, *Mater. Today*, 7, 26 (2004).

Wang, Z. L., Towards self-powered nanosystems: from nanogenerators to nanopiezotronics, *Adv. Funct. Mater.*, 18, 3553 (2008).

Yu, M., Long, Y.-Z., Sun, B., Fan, Z., Recent advances in solar cells based on one-dimensional nanostructure arrays, *Nanoscale*, 4, 2783 (2012).

Zhang, Y., Liu, Y., Wang, Z. L., Fundamental theory of piezotronics, *Adv. Mater.*, 23(27), 3004–3013 (2011).

Ziman, M., *Principles of the Theory of Solids*, Cambridge University Press, Cambridge (1979).

Žutić, I., Fabian, J., Das Sarma S., Spintronics: fundamentals and applications, *Rev. Mod. Phys.*, 76, 323 (2004).

## Website References

http://commons.wikimedia.org/wiki/File%3ATransistor_Count_and_Moore's_Law_-_2011.svg.

http://en.wikipedia.org/wiki/File:NAND_scaling_timeline.png.

http://it.wikipedia.org/wiki/File:Fullerene-C60.png.

http://it.wikipedia.org/wiki/File:MolecularImagingTherapy.jpg.

http://m. technologyreview.com.

# Index